The Internet Phone Connection

Cheryl Kirk

Osborne **McGraw-Hill**

Berkeley New York St. Louis San Francisco Auckland Bogotá Hamburg London Madrid Mexico City Milan Montreal New Delhi Panama City Paris São Paulo Singapore Sydney Tokyo Toronto

Publisher
Brandon A. Nordin

Acquisitions Editor
Megg Bonar

Project Editor
Nancy McLaughlin

Technical Editor
Dave Kosiur

Associate Editors
Cynthia Douglas
Heidi Poulin

Copy Editor
Luann Rouff

Proofreader
Stefany Otis

Indexer
Carl Wikander

Acquisitions Assistant
Gordon Hurd

Computer Designer
Leslee Bassin

Illustrator
Richard Whitaker

Series Design
Seventeenth Street Studios

Cover Design
em design

Quality Control Specialist
Joe Scuderi

Osborne/**McGraw-Hill**
2600 Tenth Street
Berkeley, California 94710
U.S.A.

For information on translations or book distributors outside the U.S.A., or to arrange bulk purchase discounts for sales promotions, premiums, or fundraisers, please contact Osborne/**McGraw-Hill** at the above address.

The Internet Phone Connection

1234567890 DOC 9987

ISBN 0-07-882269-6

To Mom, Dad, and Aunt Jane, who are always thinking good thoughts. And to Mike, the best friend a girl could ever have.

Oh, and don't let me forget Helen Orme, and the Man upstairs.

About the Author...

Cheryl Kirk is a computer consultant and journalist living in Anchorage, Alaska. An avid Internet phone user, Cheryl takes full advantage of this money-saving technology to stay in touch with her family and friends, share ideas and advice with other users, and chat with a variety of fascinating people who live and compute all over the globe.

CONTENTS

Acknowledgments

A book is not just one person's endeavor. It's a group effort, and there are so many people to thank. Megg Bonar, acquisitions editor, is first on the list. She took a chance and gave me an opportunity, and made this a truly enjoyable project to do. I simply can't thank her enough. And many thanks to Megg's assistant, Gordon Hurd (Mr. Outdoors), who kept all the paperwork flowing.

Nancy McLaughlin and Osborne's editorial staff should be applauded for finding the time and patience to plow through the text with a fine tooth comb, making sure my words made sense and looked good in print. I certainly appreciate Nancy's efforts, as well as her good humor when it came to last-minute changes. This book simply couldn't have happened without her and her hard-working associates.

My technical editor, Dave Kosiur, deserves special mention; thanks, Dave, for helping me keep the facts straight.

I'd also like to thank all the vendors who agreed to put their products on the CD-ROM. Without the new and innovative products being offered by enthusiastic vendors and programmers, the Internet wouldn't be half as fun, and this book never would have been written.

Maybe it's the stress of working day and night, and drinking too much Josta Cola, but I feel compelled to start thanking everyone—including the stock boy at the local grocery store. I know this might start to sound like an Oscar's award speech gone bad, but...definitely the people at the Anchorage Daily News deserve a big thank you, specifically Bill White, Business Editor, and the editor of my computer column. Bill has put up with me all these years; why I don't know, but I do appreciate it.

And I certainly want to thank Kathleen McCoy, Lifestyles editor, who agreed to run the piece about my folks getting online which became the foundation for this book. That one story certainly influenced a great deal of people locally, and now it lives on here to do the same at a national level.

Also thanks to Howard Weaver, Kent Pollock and McClatchy Newspapers for keeping the column going, and to all the readers of the Daily News for their support.

I'd also like to acknowledge the people who have influenced me in my computer career and encouraged me to become a nerd—especially Ken Coller, who taught me practically everything I know about computers, and Bobby Diggins, who keeps teaching me more and more every day. Plus Larry Makinson, who constantly reminds me of how exciting computers really are.

And if I didn't mention a few more friends who cheered me on and talked me into writing a book, they'd probably never speak to me again. (Hmm...now that's a thought...) Thank you Joann McBride, Diana Waddell, Kim Ferrise, Dolores Farris, Judith Brogan, Marsha Tidsnick, Nick Sinkchair, Bob Gould (for renting me this great office and putting up with me), Karen Jones (for the chocolates and leniency on the rent), and of course my financial advisor, Dave Simler. Again thank you all for your encouragement. I apologize if I've left anyone out.

I'd also like to thank my Florida Senior Citizens sales force, Aunt Jane, Ruthie and Irwin, Aunt Ro and Uncle Jack, Uncle Bob and Aunt Ilsa, and Dorothy Slung. Aunt Jane and Ruthie are living examples you're never too old to start computing.

A great big thanks to Mom and Dad, for always being there and always thinking good thoughts.

And finally, I'd like to thank you for buying this book and contributing to my computer-buying fund.

TELEPHONE

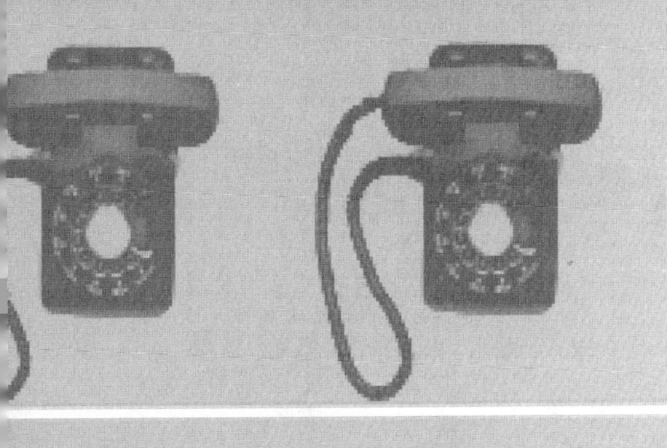

Chapter 1

An Introduction to Internet Phones

WHAT exactly is an Internet telephone? How do people use them? Where did this technology come from? How are the telephone companies reacting to people using these potential revenue-zapping programs? Is there a future for Internet telephony? Can you connect an internet phone to the regular telephone system? In this book I hope to answer all these questions.

Internet telephones are made possible by software programs that transmit your voice in real time to another person via the Internet. Today thousands of folks are talking up a storm with products such as VocalTec's Internet Phone, shown here:

Internet phones offer the opportunity not only to save planty of money on long-distance phone calls, but also to meet other people and communicate in ways never before thought of.

In this chapter, I'll give you a formal introduction to Internet telephony, including a brief glimpse of how I got started talking on the Net. I'll tell you about some of the people I've met from all over the globe, and recount their unique Internet phone success stories. You don't have to be a genius or a certified network technician to make this technology work for you. Just about anybody with a little patience, some time on their hands, and an Internet-connected computer can do it. But just in case, Chapter 9 of this book provides tips and troubleshooting advice to help you solve some common problems and get the most out of your Internet phone.

So what are we waiting for? Let's get started.

My Story

It all started when I got a $300 phone bill.

You see, I live in Alaska where the only two long-distance phone companies to choose from are expensive and more expensive. So calling the folks on a regular basis means taking out a second mortgage on the house. Then it came to me. The Internet would be the perfect way for me and Mom to discuss the finer points of clip art, or for Dad to ramble on and on about the latest long-lost relative he'd just found.

At least half the battle was already won. They were semi-computer literate. For a couple of years, Mom had been using an old Lisa (the precursor to the Mac) I had scrounged from an auction. Recently Dad had purchased a new Packard Bell 386. So they knew how to turn on a computer.

But Dad had a bad taste in his mouth when it came to online services. His initial entrance into the online world was a vain attempt at trying to navigate the then text-only CompuServe. He finally gave up in disgust saying, "I'm not going to pay by the hour just to stumble around and never find anything. It's too complicated. Talking on the phone is much easier."

So I worked on convincing Mom instead. I told her how wonderful it would be to find new clip art on the Net and give Dad something to do besides watch TV; and above all she would always be assured of getting an e-mail message from me every morning. She soon warmed to the idea, and within no time Dad found a local flat-fee Internet provider, while I found them a great deal on a more powerful Mac and a 28.8 Kbps modem.

Soon we were messaging back and forth—four, five, sometimes six, times a day. Without a doubt they were hooked. Dad was checking out the Smithsonian; Mom was looking up old high school chums. For the first time since moving away from home, my sister, cheap in her own right, was jabbering away with all of us via e-mail. That's when I decided they were ready. Ready for the big time—videoconferencing with a QuickCam camera. I had seen these little round marvels at a computer show and figured for $99 how could I go wrong? I snatched up a couple, sent one to the folks, and told them how to download the videoconferencing software, CU-SeeMe. Then we decided to meet virtually one Saturday afternoon.

Our first meeting was successful, sort of...

> *Date: Mon, May 8, 1995 19:37*
> *From: Joseph W. Kirk*
> *To: Cheryl L. Kirk*
> *Subject: Re: Just let's meet directly*
> *I've just read your message. Glad that you were able to see me, because all I got was a black screen. Perhaps I should get Mom up to sit in front of the computer since she looks better than I do.*

It was a start. After a little more encouragement and a few more instructions, Dad agreed to meet me online again the next day.

> *Date: Tues, May 9, 1995 21:22*
> *It worked!*
>
> *Date: Fri, May 12, 1995 22:28*
> *From: Joseph Kirk*
> *To: Cheryl Kirk*
> *Subject: Re: Try this...*
> *Really great to see you. This is very exciting. You are right. I really like seeing you better than talking over the phone. Hope you are not spending all your valuable time on the computer with us. Will send you another message tomorrow. Goodnight. Love, Mom.*

The looks on their faces as their picture popped onto my computer screen were nothing less than sheer amazement. Not only could we talk back and forth with e-mail, we could now see one another as well. They were both hopelessly hooked.

Unfortunately, when we tried to use the audio portion of the program, it became apparent that the Mac IIci I had bought them secondhand was woefully inadequate. Although we could see one another, the best sound we both got was an occasional burst of garbled noise. So for the time being we resorted to sending e-mail and waving at one another a lot.

Then Mom called me on the telephone one day in need of technical advice. At the end of the conversation she said, "I really do miss hearing your sweet voice; I wish there were some way we could really talk over the Internet." From that point on I became a woman with a mission. If you could send big files, if you could send video, if you could "chat" by typing messages to each other, why couldn't you speak to each other as well? I started to search the Net for a way.

It wasn't long after when I stumbled across Netphone, a Macintosh-based Internet telephone program. It worked fine when I used it on my PowerMac to talk to some guy in California, but again, when Mom tried it, the lowly Mac IIci choked out garbled mish-mash. There was no doubt, they needed a faster machine.

That was about the same time Dad purchased a Compaq 486 for work. A new program had just popped up on the Net called Internet Phone. I urged him to test it out with me, and after a few successful conversations we ended up buying the program. Soon we were talking up a storm.

Since then they've upgraded to a Pentium and installed Windows 95 to make it easier for Mom, a dyed-in-the-wool Mac user, to use. Now we talk all the time. Our conversations come over crisp and clear. On those rare occasions when I'm not on the Net, they'll leave me voicemail. They haven't picked up an actual phone to call me in months. And I can't remember the last time they left me a message on my telephone answering machine. I've saved about $200 per month on long-distance phone bills. The folks save about $80 a month. Best of all, we talk practically every day.

What Really Is an Internet Phone Program?

In the most basic terms, Internet phone programs work on computers equipped with a sound card, speakers, and a microphone. Internet telephone programs transmit voice to another computer via an Internet connection. Internet phone software turns the sounds coming from your microphone into digital signals that travel through the Internet to the other person's computer and out their speakers. It's really all about transferring data. Instead of displaying fancy World Wide Web pages like your Web browser does, Internet phone programs simply transfer, then play, the sound of your voice to another computer.

To do this, though, these programs need power, meaning the old DOS-based 386, like my Dad had been using to type letters, wouldn't work. The minimum you can really get by with is a 486/66MHz Windows 3.1-based computer. But if you are really serious, and want to make it easier, as in my Mom's case, a Pentium with Windows 95 is the way to go. Dad would have been left out of the newer incarnations of Internet phone programs that offer those nifty features like voicemail had he not upgraded to a Pentium and Windows 95.

Even though the original Internet phone programs such as Internet Phone and FreeTel started out as a way to transmit voice from one computer to another, now many of these phone programs let you leave voicemail messages, transfer files, view live video, and share programs—and this all requires a fairly powerful computer.

How Did It All Start?

Computerized transmission of audio and video has been around since 1990. However, as far as I can tell, the concept of sending voice over the Net with a *personal computer* started at the University of Illinois when a communications architect named Charlie Kline, as he says in his Web page,

> ...got annoyed at the Fall 1992 Internet Engineering Task Force meeting when I was told that the "only serious platform" for multimedia conferencing was a hefty Unix workstation. I figured a Macintosh has better audio processing ability than a Sun (true!), so I set about to write an audio conferencing tool for the Macintosh that would interoperate with the popular vat program for Unix.

In 1993, Kline released Maven, the first program to transmit voice over the Internet from a personal computer. About the same time, CU-SeeMe, a

Macintosh videoconferencing program first developed at Cornell University, was becoming the hot multimedia application on the Net.

In April of 1994, during the flight of NASA's shuttle *Endeavor*, NASA transmitted live video coverage of the space shuttle with CU-SeeMe, while experimenting with the transmission of audio via Kline's Maven software. Audio received from Lewis Research Center was then fed into a Mac, which in turn used Maven to retransmit the sounds of the astronauts across the Internet for anybody to hear.

Because Maven worked so well with CU-SeeMe, offering audio to complement the video, it was eventually incorporated into CU-SeeMe. Now a fully functioning audio and video version of CU-SeeMe is available for both Mac and PC platforms.

Around February of 1995, a small Israeli company called VocalTec offered its first beta version of Internet Phone, a Windows-based program for those fortunate enough to have multimedia-equipped PCs.

VocalTec had hoped to use the wildly popular Internet Relay (text-only) Chat channels to act as a two-way transmitter/people finder. Great idea when you think about it. You find a chat channel that interests you, fire up Internet Phone, and start talking to people with similar interests. Unfortunately for VocalTec, it failed to contact the organization that runs the IRC, the Eris Free Network, or EFNet, to let them know of the potential for increased traffic, and Internet Phone was quickly banned from public IRC channels.

It took a few weeks before VocalTec resolved its differences with EFNet, but shortly thereafter a private Internet Phone server network was created. Since then, thousands of people have downloaded Internet Phone from VocalTec's home page (http://www.vocaltec.com) and are meeting right now, talking to one another from literally every point on the globe.

It didn't take long for other companies to realize the potential of people being able to talk to one another half-way across the world without having to pay for a long-distance call. During the fall of 1995 a flurry of Internet telephone products started hitting the market.

In September of that same year, one such product, DigiPhone, produced by a small company in Dallas, Texas, became not only the first Internet telephone product to hit computer retailers' shelves, but also the first to offer "full-duplex" capabilities, which allow users of DigiPhone to talk and listen at the same time.

Soon people like Sam Diab, of Dallas, Texas, were organizing marathon "talk sessions" using products like DigiPhone, whereby people from all over the country connect and talk for hours on end.

Today's Internet phone programs have become so popular it's sometimes hard to get on popular servers simply because people across the globe, from every state in the U.S., to Italy, Russia, the Netherlands, Pakistan, Brazil, Australia, and all points in between are chatting up a storm.

The Future

Let's face it. Computers have always been tools for communicating. You write a letter with a word processor; you're communicating. You create a pie chart from a spreadsheet; you're communicating. You keep track of data in a database then create reports for the boss; again, you are communicating. So it's only natural that the computer slowly takes over the role of the telephone. Both are tools for communicating, but the computer can do it so much better by bringing together all the pieces of modern communication in one package.

It's just a matter of time before you see the computer replace the telephone as people start to use the Internet more and more for communicating. Eventually you won't need two phone lines in your home or office—one for the telephone, one for the computer. Instead, your Internet connection will become your phone.

Already many large companies are managing their telephone systems via a networked computer. So integration of computer and telephone is bound to happen. I think even more so than computer and television. Soon you'll see software that reliably links networked computers to phone systems so that people can make calls to people who are still using regular telephones. And instead of installing a phone *and* computer at everyone's desk, there will only be a computer. Telephone voicemail systems will be replaced with Internet telephone voicemail systems.

On the consumer end, the computer won't replace telephones per se in the near future, since the horsepower needed to run today's Internet telephones is still pretty hefty. But you will see Internet phone products prepackaged with many newer systems, just as CD-ROM systems that play audio CDs have become standard issue on every new computer system sold. Undoubtedly you'll see the products outlined in this book improve ten-fold in a year. And as competition becomes fierce and many of the big software houses include Internet telephony as part of their operating systems, unfortunately you'll see many of the smaller companies fall by the wayside.

What Are People Doing with Internet Phones?

Besides saving money, many people have put Internet telephony to good use in a variety of ways. I've been hearing their stories (literally) over the Net waves. Here are some of my favorites:

Brush Up on Your Italian

Julie, a 14-year-old from Arizona, told me one night during an Internet phone chat:

> I'm brushing up on my Italian. I'm taking it in school and it really helps to talk to real Italian-speaking people. I've contacted three different people from Italy. I tell them, "OK, like let's just talk entirely in Italian. No English." And they do. I understand about every third word. When I don't understand them, I tell them so in English and they help me out. It's really fun to try out what you're learning in school on people who really speak the language.

Or Why Not Your Russian?

During a recent chat using WebPhone, a schoolteacher in California told me of her successful experiment finding people to chat with in Russia:

> I teach a high school Russian class, and got the bright idea one day to try it out in class. I told the kids we would just experiment with it and see if we could find any Russian-speaking people. Can you believe on my first try I found a guy in Magadan who also spoke passable English? The kids really got excited. We must have spent the entire period talking to this poor man. He was great, though. The kids asked him all sorts of questions about what it was like to live in Russia, and did it entirely with the Russian they had learned!
>
> You can't count on the quality to be so good, and I really think I lucked out that time to find somebody who was willing to talk that long. But I just thought, Wow! What a great learning tool.

Pull Up a Microphone and Pour Yourself a Cup of Joe

"Hello, Cheryl, and welcome to Internet Cafe of Kingston, Ontario, Canada," blared my speakers one night while I cruised around the globe with Internet Phone. "We here at the cafe have outfitted four of our eight terminals with Iphone," said Lloyd Stone, President of Internet City, Inc.

> Our customers are amazed when I give them demos. I have several customers who use it on a regular basis to prearrange a time to meet relatives in distant places to enjoy an hour of chat. Of course our dedicated 56K connection helps quite a bit. A few have actually

> made voice pals, like pen pals, you know. I'd guess about 30 percent of our customers use this feature when they rent time on our terminals.

Expanding Horizons and Markets

"I run Independent Study out of Ponte Vedra Beach, Florida," Pablo Baques told me one afternoon while chatting with Iphone:

> We provide distance education counseling, research for academic and commercial information, and search for athletic scholarships at U.S. universities for qualified students from overseas. Using Iphone, I've bumped into users from Ecuador, Brazil, and Colombia who have become interested in what our company has to offer. We now talk with them on a regular basis, discussing ways for our company to expand into those markets.

Relaxing in Bermuda While Running a Business in Canada

"Hi, Cheryl, this is Gordon Proulx here in Bermuda," the speakers bellowed one morning while I was using Internet Phone:

> I'm semi-retired, but still active in my business, Quebus, a computer firm in Canada. I find it necessary to talk to my partners in Toronto on a fairly regular basis. I don't know if you know anything about the phone rates in Bermuda, but they are high. The phone company charges $1.46 a minute for calls to Canada. On average I spend about 12 hours per month talking to my partners. Therefore, my savings by using Iphone are approximately $1000 per month.
>
> I just e-mail the secretary and tell her when I'm going to call and who I want to speak with and we normally work things out very well. I can talk to as many people in the office as I need to for as long as I like. I've even sat in on meetings with the thing running, although it's a little tough to hear sometimes. Otherwise, just talking to people in general is very entertaining, more so than sitting in front of the television. In fact, I have had numerous discussions with people on Iphone about how one emigrates to Bermuda!

Could You Please Pass the File?

"Well, I do do my fair share of wasting time on this thing," said John Silver, a financial analyst for a company in North Carolina, one night while we chatted using FreeTel:

But I really do use it for business. I have to send financial reports to our home office in California. We tried just putting them on the LAN server, but found that people who weren't supposed to be reading them had access to them. Plus, I'd always get calls from my counterpart in California asking me, "What about this entry?" or "Are you sure about this figure?"

So when I ran across FreeTel one night, I realized this might just work for our situation. We tried it just to check out the voice quality and we found no real problems other than the fact we couldn't both talk at the same time. Our next step involved transferring the daily spreadsheet file. Well, it came right to Susan. She looked at it and was able to ask me questions about it right then and there. I dunno what kind of money we've saved the company on long-distance charges but I bet it's substantial. We've been doing this for a couple months now with no real problems. It sure beats trying to teach somebody how to ftp.

note *FTP stands for file transfer protocol, a way to send and receive computer files on the Internet. When you "FTP" a file, you are sending that file to a server or main computer repository. Sending files, or FTPing as it's called requires a special program called a file transfer program, that lets you connect to the other computer, move to the directory or folder where you want to put your file or files, and then send or transmit those files to the other computer.*

Get Cheap Internet Access—Call Me to Find Out How

While cruising around looking for someone to chat with, I stumbled across a user whose name was "Get Cheap Internet Access." I was intrigued and rang him up. But instead of a live person answering the phone, I was amazed to hear a prerecorded message telling me all about how I could get Internet access for $15 a month. Prerecorded voicemail messages advertising a business. How clever, I thought... Then I ran into a car dealership that was experimenting with including pointers to WebPhone from their home page so people could "call" and hear what specials were running that day. "We know it's a small spattering of people who have this technology, but we know there's a future," an e-mail message replied when I inquired as to the company's success.

Best Demonstration of a Company Really Using Its Product

Definitely this award would go to NetSpeak, since they use their own WebPhone product to answer technical support questions. It's so easy to dial them up. No need to find their Web page or hunt for their 800 number. Tech support is just a click away if you are using WebPhone.

Checkmate

"I have to admit I use it for something totally nutty. I play chess with a fellow in South Africa," Don Nelson, from Florida told me one night while testing out WebPhone:

> It may sound strange, but I met this guy just testing out the program. We got to talking and sure enough he's a real chess buff. Only problem is neither of us had any regular chess partners to challenge us. So we got to talking and figured out we could play each other. Each of us would just set up a board, then we would call out the moves to each other over the WebPhone. He's beaten me half a dozen times, and I've done the same to him. It's really pretty silly, I know, but it keeps me busy.

Houston, We Have a Problem

NASA regularly broadcasts shuttle missions via its CU-SeeMe reflector. CU-SeeMe is more like a videophone than a regular phone, offering participants the ability to see one another and, in some cases, like NASA's case, hear what's going on as well. "Our class has been watching the shuttle missions," a teacher from Oregon told me one day via the chatting/typing feature of CU-SeeMe:

> We've spent many a day watching astronauts work computers, dock with other spacecraft, and float around. It's really exciting for the kids to see it happen live.

Sex, Lies, and FreeVue

"I'm telling you these people get on all hours of the day and night and have cybersex," my friend Bobby told me time and time again. "I was on one CU-SeeMe reflector site, and you could see everything. I mean everything." It's true. Increasingly people are beginning to use these products as they would 900 numbers. I had never run into this type of thing until trying out FreeVue one night for the first time.

FreeVue is a mix between an Internet telephone and videophone. You can hear, type, and see the other user granted he or she has a camera. I thought I would get a glimpse of the late night geeks that frequent the Internet phone servers. Instead I got a whole lot more. Suffice it to say these were head shots of a different kind. As with anything on the Internet, I strongly advise parents to monitor the use of any videophone-type program.

Let's Get Together and Talk About Goats

Earl and Claudia Fitzgerald of Silsbee, Texas, raise exotic pigmy goats. "They're show goats—you know, like show dogs, only with a higher class of people," Earl told me one Sunday afternoon when I was trying out TeleVox:

> There's quite a market for show goats all over the world, especially in California. And there's quite a number of breeders, but we're all scattered across the globe. We're on a goat breeders mailing list, and that's where we got the idea we could use something like TeleVox to create a group where all of us could actually call one another and talk.
>
> Eventually we hope to offer a Web page, maybe even link it directly to the TeleVox phone group we've set up so people can call us and place orders for our goats.

We're Paying Lots for College; Why Pay for Phone Bills?

"My daughter went to Seton Hall last year and the phone bills practically killed us," says Allen Stewart of Maryland:

> I figured this Intel Phone would be a great way to save some money. She was always calling her mother last year, and the phone bills were almost as much as the college tuition. I'm testing this out to make sure everything is set up right, then I'll load up her computer with the same software and she'll be ready to go.

Believe it or not, she's pretty excited about it too, since a lot of her friends are getting into the Net. She figures she'll be able to get a few of them to get connected with Net phone business, and then she can stay in touch with them, even though they are going to different schools. This is the best thing that's come along since...well...ever!

Chapter 2

The History and Future of Internet Phones

ABOUT eleven years ago I met a pimply-faced kid, I'll call him Jerry the Dweeb, a nerd definitely ahead of his time. I'd describe him as one of those typical young geeks lacking in all the social graces, including an understanding of how important personal hygiene really is. But aside from his social shortcomings, he had unique insight into the future of telecommunications, an insight I don't think anyone understood at the time, including me.

Jerry's dream was to create a computer telecommunications program that would link two computers together not only to exchange text, which at the time was about all most telecommunications programs could do, but also to transmit voice.

I remember one of my friends laughing at this idea when Jerry detailed the particulars during a local computer users group meeting. In those days, most people telecommunicated through computer bulletin boards—text-based systems running on individual computers, which usually connected one or two users.

Most of us who were frequent users of bulletin boards prided ourselves on having two telephone lines installed—one for data communications and one for voice. So the need for simultaneous data and voice transmission wasn't there, or so it seemed.

"What a dumb idea," my friend said to Jerry when he explained how his latest greatest invention was going to work. "Why would anyone want to do that when you can just pick up the other phone? What a silly idea. You're wasting your time. And besides, modems will never be fast enough to carry both voice *and* data."

Little did we know that geeky Jerry really was onto something and way ahead of his time. When advancements in modem technology boosted transmission rates, when the government turned over control of the Internet to corporate America, and when the media started hyping the

Net and propelled millions of people to get on it, soon it was clear that his idea really made sense.

For me, Jerry will always be a good example of how some computer users even as far back as 1985 were thinking, "Hey, maybe there's a way to put voice and data on the same line!"

The Start of It All

About 1992, when the Internet first began touting a graphical interface and more and more people found it easier to get connected, the first multimedia voice and video broadcast took place. The Internet Engineering Task Force held its first meeting using the relatively new MultiCast Backbone portion of the Internet to broadcast the event.

note Multicast *basically means sending information to more than one person at a time. Think of the* MultiCast Backbone *(or the* MBone*) as a special channel on the Internet specially made for sending big chunks of data like sound and video to many people.*

Specific videoconferencing software used over the MultiCast Backbone lets individuals and groups communicate with each other using sound and video. University and government agency networks, often referred to as *subnetworks*, have high-speed links to the Internet, and use the MBone to videoconference with other universities and agencies across the globe.

This combination of virtual network and custom software, designed to support the routing of multicast data packets (packages of voice and data wrapped up into computer bits and bytes that are sent to more than one recipient) between networks, was really the first true Internet telephone system. It continues to offer users of direct, high-speed connections the ability to chat, see each other, and transmit computer data all at the same time.

Technology is never happy to sit idle. As more people wanted to use their desktop computers and modems to connect to the Internet, innovative people like Charlie Kline, the inventor of Maven who you read about in Chapter 1, figured out a way of programming a standard Mac to use an Internet connection for transferring voice.

Kline, a real Mac-lover, realized that if a high-powered workstation not meant for sound could still transfer multimedia to numerous users, then the Mac—the perfect multimedia computer, could most likely do the same, even if it meant transferring the sound to only one or a few listeners. The MBone spurred Kline on to create a program for the rest of us, non-MBone connected users.

After the successful demonstration of Maven and CU-SeeMe at NASA, the stage was set. More and more people realized that if you could transfer data, you could transfer sounds, and soon companies like VocalTec started to spring up all over.

People around the world, who may have had little in common other than the software programs they were using, started talking up a storm, forming electronic friendships, exchanging advice on computer hardware and software, learning about different cultures, and communicating for hours on end free from worry that in a month or so their telephone bills would finally catch up with them. Internet phones have started to really take off, and in the next year you should see tremendous growth in the number of software, hardware, and Internet service providers catering to Internet telephony.

Telephony Today

When you stop and think about it, the world of Internet telephony has already begun to explode in just a very short year and a half. Today no fewer than a dozen companies offer Internet telephone products with all sorts of features not found in the first version of Iphone, such as voicemail, collaborative whiteboards (electronic sketch pads that two or more people can share to draw pictures, type text, and sketch diagrams), and the ability to share applications (such as your word processor to enable the other person to read what's on your screen, or a spreadsheet so you can compare financial figures, or maybe your favorite game of Hearts).

Emerging Standards

Currently, instead of added features, interoperability between competing Internet phone programs is the big buzz. Up until now, if one person purchased Internet Phone and the other purchased DigiPhone, they simply couldn't talk to each other via computer. Chronic non-compatibility was bogging down the Internet telephone world, dividing people into disjointed electronic islands according to the software they happened to own.

So Intel, with its Internet Phone, and Microsoft, with its NetMeeting, decided what the hell, why not take advantage of a set of standards a consortium of Internet phone-related companies called the International Telecommunications Union had developed for connecting and transmitting sound? The ITU's H.323 Recommendation, which is a set of standard compression schemes or instructions on how to compress sound into small data packages, let different vendors of Internet phone programs, terminals, and equipment—such as hardware-based videophones—operate with one another. When a program follows the ITU's standard set of instructions, it can operate with other programs that do the same.

Obviously, by deciding on standard compression schemes or ways in which people with different products could talk to one another, they understood that this could only help consumers. One day all Internet phone products will use this standard, much like traditional telephones use a standard compression scheme today. That way, if you decide you'd like to have the voicemail features offered by WebPhone, and your cousin would rather try out those nifty sound effects in TeleVox, you'll still be able to communicate with each other.

Still, There's a Long Way to Go

By far, though, the most significant—and needed—enhancement to Internet phone technology in the last year has been the improvement of sound quality. What was a very broken, sometimes inaudible, conversation in the early 90s has progressed to a relatively clear, albeit slow, CB-type mode of communication.

Yet there is no doubt: Internet phones still have a long way to go before they rival standard telephones in terms of reliability, voice clarity, and ease of use. Although I use the products on a regular basis to chit-chat with Mom and Dad in Houston, it does take a great deal of patience, time, and energy. Sometimes I'll have computer problems; sometimes Dad will. Sometimes the Net is so busy, we both sound like we're being bounced around in a tin can. Oftentimes, its just the luck of the draw as to whether we can hear each other well enough to carry on a normal conversation.

As computer programmers figure out new ways to compress sound into even smaller packages and compensate for Internet delays, you'll see (and hear) the sound quality improve. And as more and more people upgrade their computers to multimedia systems, more and more people will be accessible via Internet phones. Like every great invention, from the television to the radio, it takes time for people to catch up with technology, and technology to catch up with people's wants and needs.

Internet Telephones: Almost as Good as the Real Thing

Although I've had some really fantastic conversations with people as far away as South Africa, I've also had some horribly bad connections with people as close as Seattle. Some of these problems can be attributed to the people running the software at the other end. Others are the result of competing Internet traffic, strangely behaving beta software, and even huge sound files being forced down the pipes of networks that were originally designed to carry simple data. The simple fact is, Internet telephones are only *almost* there.

Even so, between my folks and me, we've saved several thousand dollars over the past year and a half that we've been using Internet phone software. But I have to be honest here. We've also spent about a thousand together on Internet access, plus another couple of thousand on computer equipment—not to mention the cost of some traditional long-distance calls to help Dad set things up. In reality, all in all we've come out about even.

Obviously, we haven't trash-compacted our real telephones just yet, forever convinced that Internet telephones are the be-all and end-all. From time to time I still pick up the phone to let Mom know of the latest fire, flood, or earthquake that's hit the Anchorage area.

Face it. The telephone provides people with a tool for immediate communication. It will be a long time before every person on the planet converts from using the telephone to using the Internet for their sole means of communications. Simple economics dictate that right now, since an estimated 60 percent of Internet users are connected via dial-up connections and most people aren't on the Internet *all* the time. You can't just click someone's name in your Internet telephone program and expect them to answer. They may not be there. You still have to plan, e-mail, and schedule online conversations with the people you really want to talk to.

And when you look at the overall, bigger picture, only a small percentage of people who have telephones have Internet access. And of those who do, most can't afford the several hundred dollars a month it costs to stay connected to the Net all the time.

It will take definitely take time before you see everyone routinely using Internet phones. But in the meantime, many innovative telephone companies have decided to start taking advantage of the Internet and its ability to connect people throughout the globe cheaply and easily. Companies such as the Maryland-based Labs of Advanced Technology Corporation are starting to connect telephone callers together via the Internet instead of through the telephone networks. The concept is relatively simple: The company sets up a computer connected to the Internet and the local telephone network on one end, and another Internet/telephone network-connected computer on the other end. People then use their normal telephone to dial this local Net/phone-connected computer. When a caller enters the long-distance number, the computer sends the call over the Internet instead of routing it over standard long-distance lines, thus avoiding the long-distance charges and fees. Look for more and more of these types of companies to spring up as people begin to realize that there is a cheaper way to make phone calls.

ACTA and Their Petition

That's why I've never understood why an organization called America's Carriers Telecommunication Association, or ACTA, filed a petition in March of 1995 with the Federal Communications Commission, stating that

> ...the providers of this Internet Telephony software are telecommunications carriers and, as such, should be subject to FCC regulation like all telecommunications carriers.

Who is ACTA? It's a trade group representing 130 small regional long-distance carriers or companies that buy "time" from the big telephone companies and resell it to the consumer. Since they don't own the telephone and network wires that make up the Internet but simply resell airtime, so to speak, these companies are threatened by any service that could very well eliminate them altogether.

According to ACTA,

> ...it is not in the public interest to permit long-distance service to be given away, depriving those who must maintain the telecommunications infrastructure of the revenue to do so.

But ACTA is flat wrong in their position. The telephone companies, not the resellers of airtime (the companies that make up ACTA), are the ones that maintain and pay for the telephone infrastructure. Telephone companies, not resellers, actually carry the compressed Internet telephone voice files across *their* telephone wires as well, and pay to maintain these lines. And that's probably why MCI and AT&T, two of the major players who provide the Internet backbone that makes Internet telephony possible, don't back ACTA's position.

The members of ACTA would lose big if people started using the Internet to make long-distance calls. The big phone companies stand to make money since they own the wiring that makes up the Internet, but resellers of airtime wouldn't since there would be no need for them anymore. Basically, ACTA wants the FCC to halt the sale of Internet telephone programs, hoping to invoke regulations that would restrict voice transmission across the Internet.

What they obviously want to block is competition: software manufacturers that are rivals. Rivals they'll never be able to compete with, because the regional carriers don't own anything—not the wires, servers, or software—

nor do they produce anything. All they do is resell time. Internet phones don't benefit them in any way since they can't resell any portion of the Internet phone equation. Instead of changing with the times, this group of resellers has requested that the government place restrictions on what is essentially free trade.

Silly, huh? Thank goodness for us, the consumers, that the large phone companies don't see things the same way ACTA does.

Don't Include Us!

A number of phone giants, including AT&T and MCI, have gone on record saying they won't be a party to ACTA's petition. Their reason? These larger companies are in favor of letting the new technologies develop without interference. Why? Well, Internet users are still paying for their services via their Internet providers, who in turn are paying for leased lines. In essence, the revenue still comes back to the phone companies. It's just a matter of moving that money from one network carrier back to another.

Wisely, the FCC has decided, at least for now, not to intervene. In the summer of 1996, soon after the FCC's decision, several major players such as Microsoft and Intel announced plans to incorporate Internet phone technology into their upcoming products. This development helps ensure the future of evolving Internet telephone technologies.

The Future...and You

I don't think there is any doubt: The future of Internet telephony is pretty darn rosy, even if there are some naysayers and poor sports nipping at the heels of companies that manufacture the software.

Certainly as more and more people discover the Net, and as more and more people realize that they can now talk for an unlimited amount of time to anyone across the globe, their excitement will spread like wildfire. The increased demand will mean more competition, which most likely means lower prices. To handle the increase in users, Internet service providers will need more and more bandwidth from those large telephone companies. Selling more bandwidth means more control over revenue. Ultimately, the big phone companies will be able to control prices and will benefit directly.

Here's a story that should give you an example of how popular Internet telephony has been even in its infancy: In July of 1996, when Intel unveiled their Internet Phone product, they announced that it required at least a 90MHz Pentium. In less than eight hours, over 2000 people had downloaded and registered the program.

There is no doubt. Just like my family, other consumers are excited about the prospect of saving money. And if that means having to upgrade the home computer to save a few bucks, so be it. When people do decide to upgrade, companies like Compaq, IBM, and Packard Bell will be waiting for them, offering advanced multimedia capabilities such as built-in sound cards, speakers, and microphones as standard fare. Some manufacturers are even packaging their machines with demonstration versions of Internet telephone products. Major computer manufacturers realize that Internet telephony is yet another way of enticing consumers to buy their products.

The Corporate World Will Demand This Technology

The most motivating push to develop Internet telephony is coming from corporate America—and that push is growing more and more urgent.

Most large corporations already use the Internet to transfer company data; however, voice communications between corporate offices are usually handled through the telephone network. But consider how much money a company could save if it no longer had to pay traditional long-distance prices for intercompany calls. Then consider how much less support would be needed if all communications were handled on one network. There would be no need for separate telecommunications departments. It's really in the corporate world where Internet telephones will blaze new trails.

Already VocalTec, Integrated Device Technology Inc (IDT), and Global Exchange Carrier (GXC) are leading the way with gateways that link traditional telephone service and Internet phone software together.

With VocalTec's Internet Telephone Gateway products, business travelers can use their PCs to connect to the Internet and get their voicemail, even though it may be stored on a voicemail system miles away. No long-distance charges—just a Web browser, an Internet phone program, and an Internet connection.

IDT and GXC are taking a slightly different route by offering discounted long-distance service for users who would rather use their computers than their standard telephones to call other people worldwide—even if those other people don't have Internet phones. Still other products, such as Amail, offer standard telephone users the ability to check their e-mail messages from any telephone system.

My Predictions

Of course, all these developments signal the marriage of the telephone and the computer. As I circle round my crystal ball, I see long-distance telephone companies moving away from the consumer market, opting instead to provide "backroom" connections to Internet service providers. Less fuss. No muss.

And I see Internet service providers stepping up to the plate to offer higher and more reliable bandwidth to those who have been demanding it. I also see each user's e-mail and voicemail box becoming one. With Internet-to-telephone gateways, the merging of e-mail and voicemail is now possible—and it will become increasingly prevalent. People will soon realize that having to deal with all these different boxes—e-mail, voicemail, snail mail, and fax mail—is just too much.

I see companies like ARCO Alaska deciding to upgrade all their desktop computers to videoconference-capable machines. And with the integration of NetMeeting into Microsoft Office 97, I see companies like Microsoft making it easier for corporate clients to integrate voice and application-sharing collaboration into the way companies do business.

I also see small and large businesses jumping on the bandwagon, offering all-in-one customer support and the ability to place orders over an Internet phone right from the company's Web page. Instead of having a separate 800 number, companies will use a combination of their Web pages and Internet phones to offer voice communications via clickable icons found on their Web pages. These icons will bring up Internet telephones and connect the person browsing the Web to the customer support hotline or purchasing desk. It's already beginning to happen on a limited basis. Watch this type of advanced communication explode as Internet telephone companies begin to offer advanced all-in-one voice communications integrated into Internet services for all sizes of businesses.

Already, innovative companies such as Netphonics, with their Web-On-Call program, offer the ability to have Web pages read to a telephone caller via a computer that converts text to speech and reads it back over standard telephone lines.

Finally, I see more and more individuals like you and me using Internet telephones to reach out and touch their friends and family. As Internet phones become more and more reliable, sound quality gets better, and the interfaces are made easier to use, it's a no-brainer. Internet telephony is gonna be big.

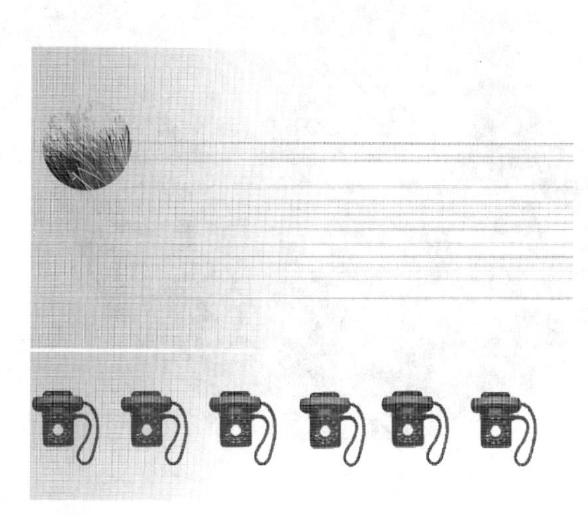

TELEPHONE

Chapter 3

Building Your Internet Phone: The Components You'll Need

WHEN you want to make a *real* telephone call, the only hardware you need is the phone itself. But an Internet phone call is a little more complicated, and without a doubt requires a fair bit of patience and a tad more hardware.

Whether you're using a Windows-compatible machine or a Mac, you'll need some basic pieces—parts that may not have come with your computer when you originally purchased it. Let's take a look at exactly what you'll need and how to put it together. In this chapter, I'll outline the major components needed to work with the wide variety of Internet phone programs. There are a ton of sound cards, a mountain of speakers, and a wide range of microphones to choose from. I can't cover every manufacturer, but I can tell you what to look for.

Getting Started

What do you need to get started? If you already have a computer, what upgrades should you buy to get crisp, clear sound quality? What about laptops? Can you modify them so they have sound and speech capabilities? And which computer should you get if you don't have one? Do you buy a Mac or a PC? A 486 or a Pentium? PowerMac or Performa?

The answer? Get the fastest computer possible with the best sound card and most expensive microphone made. There. That's all you need. End of chapter.

Wouldn't it be great if it were that simple? I've always agreed with the computer genius, Steve Jobs, that a computer should be an appliance—easy to use, easy to configure. Just like a toaster. The computer world is too technical. Every time some nerd wants to engage me in technospeak about the latest 32-bit, 97MHz, 28.8 baud, 256K cached whizbang, my eyes glaze over.

Buying the Right Computer

My approach to recommending or buying computer hardware is very simple. Buy the best within your price range—and the best usually means the biggest numbered thingie-madoodle. Above all, don't agonize over the purchase. As long as whatever you buy meets the minimum requirements, you can be sure you'll have reasonably good Internet phone conversations.

Today you can get a decent Pentium (also called a 586 on some manufacturers' spec sheets) for under $1500—which, by the way, is about the amount I saved the first year I started using the various Internet phone programs.

So if you're just starting out, and you're interested in talking mainly to family, I recommend that you gather your phone bills for the past year to see how much you've spent talking to aunt Jane, sister Charlotte, or nephew Ryan. Then consider if your relatives also have a computer with enough horsepower, and the time and patience to use Internet telephones to talk to you. In my case, I started by examining the phone bills to my parents.

Over a year's time, I was spending from $3000–$4000 talking to just Mom and Dad. I decided that if I talk to them about 50 to 80 percent of the time through an Internet telephone, I could use that as a basis to budget for a new system for myself, and maybe even use some of the extra savings to help Dad upgrade to a Pentium computer. Within a year's time that idea became a reality and I had not only saved enough money to buy both Dad and myself a new Pentium computer, but enough for a color printer as well.

My sisters were another story. Their computer needs were not as extensive as my folks', but their equipment was much older and not powerful enough to use even the lowest-end Internet telephone. But they realized that even if they couldn't really use Internet telephones, e-mail could provide a relatively low-cost form of communication that could save them some money. And by saving that way, they eventually would be able to upgrade to more expansive computers, eventually joining Mom and me in some audio chats.

Cynthia and Charlotte have been on the Net for about eight months now, and are already halfway there. Within a year's time they should have enough saved to upgrade to a multimedia computer, fully Internet telephone capable.

Their plans are to spend no more than $1500 on a complete system. I think they are pretty smart to limit themselves like this, and I'm not just saying this because they are my sisters. As history has shown, prices for computers are always dropping. Those high-priced computers today will be much less next year, sometimes as much as 50 percent less. Consider too that companies such as Wyse are considering putting microphones and speakers into their $500-and-under "Internet Terminals." So my recommendation is *not* to spend thousands on a computer. Instead buy a moderately priced computer that can be easily upgraded. Computer prices rarely go up, and technology never moves backwards. Play it conservative, match what you think you might save in one year talking to just one relative, friend, or business associate and then budget accordingly.

If you don't have a computer and are thinking of buying one, I recommend you purchase a Windows 95-based computer. Now, I love Macs. I've owned almost every major model of Mac imaginable, from the Lisa (which was the precursor to the Mac) to the top-of-the-line PowerMac. The Mac's interface is far more elegant than Windows 95. And the fact that all Macs come with built-in sound capabilities—some with microphones, and all with small, but adequate, built-in speakers—would lead you to believe that the Mac is the way to go in the Internet telephone world. But because there are more installed Windows-based PCs out there than Macs, realistic economics dictate buying a Windows 95-compatible computer. More Windows-based computers in people's hands means more Windows software on the market—in other words, more options to choose from.

A look at the list of Internet phone programs supports that opinion. So far, five times more Internet phone programs have been created for the PC than for the Mac. Sure, not all of them are great, but it's nice to have so many options—especially when Internet phone servers are busy. When you can't connect with one program, you can always use another. Plus, with so many to choose from, you're bound to find one that offers just the right mix of fancy features, such as compression rates that best match your particular machine and Internet connection, voicemail, or advanced search capabilities.

In addition, Microsoft has jumped into the ring by offering a program called NetMeeting, currently only available on the Windows 95 platform. NetMeeting, a combination Internet telephone and text conferencing program, is also an optional add-on to Microsoft's Internet Explorer browser, and will soon become an integral feature of Office 97, the combination word processor, spreadsheet, database, and presentation program. Since both NetMeeting and Office 97 will be optimized to work on Windows 95, you can bet that Microsoft will push for Windows 95 to be the de facto platform for Internet telephony.

for mac users *I'm not totally discounting the Mac here. Some of the Mac programs, such as VocalTec's Internet Phone, are good products that work well on Quadras and PowerMacs. If you need to buy a Mac, or you already have a Quadra or PowerMac, then you should probably skip this section and move on to the speaker section.*

So what type of Windows 95-compatible computer would I recommend? Although on average most Internet telephone software manufacturers recommend a minimum of a 486/33MHz computer, I'd get at least a 133MHz Pentium. Remember, the speedier the computer, the better the sound quality—and the faster the computer will record, and then transmit, your voice. This all means less delay in the conversation.

Computing Power

Your computer needs a fast engine to compress and transmit your voice through the Internet to the person listening on the other end. The faster the engine, the better your voice sounds and the quicker you will hear the other person. The computer's engine is called the *microprocessor*, sometimes referred to as the processor *chip*. Computer engines or microprocessors come in a variety of types and a variety of horsepowers.

On the Windows-based computers, the most common chip types are 386, 486, and Pentium. On the Macintosh, chip types are 68030, 68040, and PowerPC. The higher the number, the faster the machine and the more data the computer can process at one time. Pentium and PowerPCs are the top-of-the-line chips and run very fast. 68030 and 386 chips are considered low-end, slow microprocessor chips. The reason for the different numbers? Just like car companies, each year chip manufacturers come out with new models.

And just like cars, each type of chip can run at different speeds. The speed of a microprocessor chip is measured in megahertz clock speed, abbreviated as MHz. When you go computer shopping you'll most likely see numbers like these: 486/66MHz. This means that this is a 486 engine running at 66MHz clock speed. In other words, this would be a relatively fast computer. But a 166MHz Pentium processor, the next step up in terms of performance from a 486, is a real screamer, running 100MHz faster than a 486, and able to process twice the data in the same amount of time than a 486.

The first version of Internet Phone from VocalTec required only a 486/33MHz computer. The most recent version of Intel Internet Phone requires at least a 90MHz Pentium. Now 66MHz, 75MHz, 90MHz, and 100MHz Pentiums are considered outdated by today's standards. Unless you get a really good deal, don't even consider buying one of these machines, since the price difference between two consecutive speeds is almost negligible. A little comparison shopping at my local computer store showed less than $100 difference between two brand-name systems, one with 100MHz and the other with the higher-powered 133MHz processor. 100MHz is really the low-end of the computer spectrum (at least it is as I'm writing this book), and as software makers add more features to Internet phone programs, still more horsepower will be required.

remember *If the person you are talking to has a 486/33MHz machine and you have a Pentium, the 486 will dictate the speed of the conversation. The lowest common denominator is what powers any Internet telephone conversation. So if you're like me, and plan on talking to Mom and Dad with your Pentium, they may need to upgrade their 486 computer.*

But Can't I Just Upgrade?

*What do you do if you've already poured a pile of money into a computer and don't want to spend a fortune buying a new one? The processor (which is effectively the engine) of most models of computers can be upgraded from 286 to 486, and from 386 to 486, or from 486 to Pentium chips. Companies such as Cyrix (**http://www.cyrix.com** or 1-800-462-9749), Evergreen Technologies (**http://www.evertech.com** or 1-541-757-0934), and improveIT Technologies (**http://www.insight.com** or 1-800-848-9441) sell processor upgrades. Most run under $200 and will work with a variety of computer manufacturers' motherboards.*

An upgrade package is basically just a microprocessor and some software that makes it work. You or your friendly computer technician can simply remove the old processor chip (on most newer computers, the processor will snap right out), and then snap in the new chip—provided you have the proper tools and you've protected your computer from possible electrical static charges. You'll most likely have to install the software that comes with the chip to make it work at its optimal speed. And you'll have to change the settings in your CMOS configuration as well.

Just about anyone handy with a screwdriver can do this type of upgrade, as long as they do have the proper tools and know how to use them. One caveat, though: Not all motherboards are upgradeable. The best thing to do is call the chip manufacturer before you buy an upgrade, to make sure their product will work with your machine.

Proper tools include a static mat to put the computer on, a grounding strap that you wrap around your wrist and attach to some metal object to ground yourself, a chip puller, and Phillips and flathead screwdrivers. The static mat and grounding strap prevent you from "zapping" the computer with static electricity that might be running through your body. The chip puller is used to extract the microprocessor chip without damaging it. The Phillips and flathead screwdrivers are used to open the computer casing.

Random Access Memory (RAM)

RAM is short for *random access memory.* In its most basic sense RAM can be considered the working area of your computer. To use the office analogy, if the hard disk is your file cabinet, then RAM is the top of your desk. On a small desktop, your paperwork piles up so that when you need something you have to dig for it. With a large desktop, you can spread your paperwork out and quickly find what you need. The same holds true with computers. A machine with 4 or 8 megabytes of RAM simply won't function as efficiently as one with 16 or 32 megs. The bigger your electronic desktop, the quicker your computer can load and switch to the various programs when you need them. With more RAM your machine will run faster—and be less likely to freeze up or run out of working space when you try to bring up your Web browser at the same time that you're running your favorite Internet phone program.

For years, buying RAM has been as expensive as buying gold. Many people just didn't do it. But as I write this, RAM prices have plummeted, so there's no reason to hold back anymore. And consider this: A 75MHz Pentium with 16 megabytes of RAM can actually outpace a 133MHz Pentium with 8 megabytes of RAM. Don't underpower your high-powered processor by skimping on the RAM. Load your machine up with as much memory as possible.

But What If I'm Cheap? Will I Be Left Out of the Loop?

The minimum requirement for the large majority of Internet phone programs is 8 megabytes. So, yes, you can maintain your thrifty ways and still be able to use Internet phones. Just remember, the sound quality may not be so hot, and the delay in the conversation may be maddening enough to convince you to stop pinching that penny.

Early versions of VocalTec's Internet Phone, and the current versions of FreeTel and Speak Freely, are examples of some programs that work just fine with as little as 8 megabytes of RAM.

Until I upgraded my Dad to a desktop Pentium, his 8MB, 486/33MHz laptop computer worked great using both Internet Phone and FreeTel, with very reliable and mostly clear results (although the delay in hearing my voice come over Dad's slow computer drove Mom crazy). But soon he was really left in the dust when a newcomer, WebPhone, entered the market; it required more power for its voicemail feature, something that wasn't available with FreeTel or Iphone. And that's how Dad ended up with a brand new Pentium on his very next birthday.

The Hard Drive: Your Electronic File Cabinet

You shouldn't have to worry much about hard drive space. The average Internet phone program takes from 5 to 8 megabytes of space, which is nothing in relation to the gigabyte drives that now come standard with most new computers.

remember *A gigabyte is 1024 megabytes. This means that if you install IRIS Phone, which takes around 3.5 megabytes, plus all of Microsoft Windows 95, and add in Microsoft Office for good measure, you'll still have 800 or more megabytes left over.*

Hard disk space should only be a concern if you plan to use voicemail, a feature of some Internet phone programs like IRIS Phone and WebPhone, and store the messages you receive from other callers. Computerized voicemail is usually saved in the .wav sound file format, and these files can take up enormous amounts of space because every "uh," "oh," and "mmmm" must be digitized.

If you're buying a computer for the first time, don't settle for less than 1 gigabyte of space. A year from now, when you've installed every voice program on this book's CD and saved every voicemail message you ever got, you'll thank me.

The Multimedia Equation: Sound Card, Speakers, and Microphone

If the computer's processor is the most important element of an Internet phone setup, then multimedia capability is definitely the second. A good set of speakers, a microphone that doesn't pick up background noise, and a full-duplex sound card are all the components you really need. But with so many speakers, so many sound cards, and so many inexpensive microphones to choose from, how do you decide?

When you buy a new computer, you usually don't get to decide which components come with the system; many manufacturers make those decisions for you. A new system may include speakers, sound cards, and microphones as part of a preassembled computer package.

Many computer superstores sell complete systems that include the monitor, speakers, sound card, hard disk, and modem. Many of these systems are prepackaged at the factory, meaning you may not always get to select the components you want. Because these components are labeled with the particular manufacturer's name or because they may be internal components, sometimes the computer store just can't exchange these parts (such as speakers, sound cards, or monitors) and still make a profit. Oftentimes these component pieces are not priced separately, so it is difficult for the computer store to rebate the cost of the component you want to replace with one you really want.

The best thing to do is draw up a list of the components you really want in your system, and see if the computer store can match your requirements. Check with your salesperson to see if he or she can replace a low-quality set of speakers with a better brand. Ask if they'll discount a full-duplex card when you purchase it with the system. Most preassembled computers will include a half-duplex card instead of a full-duplex card, and full-duplex cards are undoubtedly the way to go when you are buying sound cards. (But more about sound cards a little later in this chapter.) If they can't work with you on the buying end, they most likely won't be much help on the support end once you get the system home. Look for a store that is willing to work with you, but remember they may be limited on what they can offer based upon the manufacturer's model you choose.

Mail order houses, such as Gateway and Dell, will generally work with you so you can get the exact setup you want. Systems configured specifically for your needs may cost a little more, both in time and money, but they are usually worth it in the long run.

Let's take a look at each component separately to help you understand what you should get.

Speakers

Speakers are probably the easiest multimedia component to buy, since all you really have to do is listen to them. Just remember, a speaker that sounds great blasting out a bass guitar may sound too muffled when playing back someone's voice.

Most computer superstores line up the speakers they sell in a row so you can hear and compare the different brands side-by-side. Superstores like CompUSA do a great job of demonstrating how different speakers sound playing multimedia noises, voices, and music. If you aren't close to a superstore, the next-best thing is to ask your computer dealer to set up a pair, or barge into your friend's house uninvited and demand to use their computer. That always worked for me until I ran out of friends and the police started following me.

Along those lines, what sounds good to me may not sound so hot to you. Buying speakers on "specs" alone is about as good an idea as buying a television without looking at the picture. So I won't be going into detail concerning the many "music" features available. Instead, I'll concentrate on specific Internet phone-related features you should look for when buying speakers.

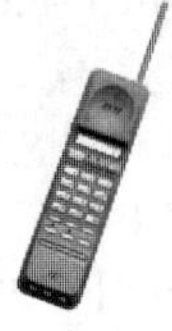

tip *Your telephone actually transmits "mono" quality voice, not stereo—so if you buy expensive speakers to use exclusively with an Internet telephone program, some of the fancy musical features will just be wasted.*

Here's a brief list of features you should pay attention to when you're shopping for your first set of speakers. (And remember, don't buy them without listening to them first!)

- **Separate AC power source** Since the typical sound card offers only 4 watts of amplification per channel, speakers need their own power to boost the sound. But I don't recommend battery-powered speakers; they are extremely limited in volume and output. If you travel with your laptop, consider a headset with a built-in mike. Less stuff to lug around. But remember to make certain the volume is turned way down on your headset before you start talking—you don't want to blast your eardrums if you connect to someone who happens to yell.
- **Clear sound when you crank up the volume** You can't base your decision on wattage alone. Some high-end speakers rate their wattage levels differently than lower-end models. Most low-end speakers are rated on their peak wattage output range, not on their ability to maintain that range. So a cheaper speaker, even one that says it compares to a high-end model, may start to crackle and sputter when the volume is turned way up for an extended period of time. This is important to remember, since some people speak with soft voices or sit far away from their microphones. Sometimes you'll need to keep the volume level cranked up for the length of your conversation just to hear the person on the other end.
- **Shielding** Speakers use magnets, and as we all know, magnets can erase floppy disks and cause monitors to go haywire. Most speakers manufactured specifically for computers are *shielded* so that the magnets inside do not affect other computer components, but this may not be the case with speakers designed for stereo systems. Be especially careful if you buy one of those mondo three-piece speaker systems. The subwoofer component in a three-piece system is usually meant to be placed under a desk, and oftentimes does not include the proper shielding for use next to a

computer or monitor. Check the technical information in your owner's manual to find out whether the subwoofer is shielded and where the manufacturer recommends placing the unit in relation to your computer equipment.

- **Easily accessible volume controls** Although software volume controls are convenient because they can be accessed through some of the Internet phone programs, they may not be accessed as quickly as manual controls in other Internet phone programs (particularly when the boss comes barging into your office wanting to know why you are waxing poetic with Petunia in Pittsburgh instead of preparing that report she needed by four o'clock).
- **Easily accessible microphone and headset jacks** Not all speakers have microphone and headset jacks, or they may be placed in awkward locations (like the back of the computer, or underneath the monitor), making them inconvenient to quickly access. If, however, you don't need to switch between headphones and speakers, or if you never change microphones, conveniently positioned connection jacks may not be very important to you. I need easily accessible connections since I'm forever unplugging and plugging things into my computer, so for me, the more accessibility, the better the product.
- **Easy setup** You don't want a jumble of wires strangling you every time you sit down at your computer. Take a close look at the back of many of the speakers on the market today. You'll find that not all manufacturers produce a neat and tidy package.
- **Reasonable price range** Most speakers that work well for Internet telephone programs run in the range of $50-$100. Beware of very low-cost speakers (in the $10-$40 range); they most likely won't come with their own power adapters, and they may not provide loud enough sound. Conversely, you don't gain that much with really expensive speakers intended for music or multimedia games—especially if they are attached to a really low-quality sound card.

remember *Be sure to match the capability of your speakers with that of your sound card. A lowly 8-bit sound card won't utilize the capabilities of $400 Bose Surround Sound speakers.*

- **Dependable warranty: repair, replacement, or refund** Most mid-range quality speakers should last you a long time. But the ability to have them repaired or replaced, or to get a refund if you aren't satisfied, is imperative. Check the store's return policy before you run off with a pair of $300 speakers under your arm. (And you really should pay for them first, you know.) If for some reason the speakers don't work, you'll appreciate being able to return them and get a full refund.

tip *What sounds great in the store may not sound so hot with your system, or in your office or home. It's a good idea to check with the salesperson to see if the store has demo units they can loan.*

Some Speakers to Consider

The following are some speakers that you might want to check out. But again, remember that what sounds good to me may not sound so hot to you—so listen, listen, listen!

note *Each of these recommendations lists the currently suggested retail price for a pair of speakers. You'll probably find the "street price" somewhat lower.*

ALTEC LANSING ACS52 ($129) A very good quality speaker with easy-to-reach volume controls.

ADVENT POWERED PARTNERS AV170 ($130) These speakers' tidy cabling and crisp quality make them a reasonable buy.

JAZZ J-590 ($100) Solid and attractive. What else could you want?

KOSS HD/100 ($100) AND NEC TECHNOLOGIES' AUDIO TOWER ($99) These are two of the best low-cost speakers on the market. They can maintain a clear sound even with the volume cranked to the max.

YAMAHA YST SERIES ($200) A more expensive speaker, but with a clean design (see Figure 3-1) and sparkling sound. Definitely works well with music and multimedia games.

remember *It is possible to use regular stereo speakers and amplifiers with your computer. A speaker doesn't have to say "computer" or "multimedia" on it just to work with a computer's sound card. But be careful where you place those "regular" speakers. Speakers have magnets in them and magnets can erase disks and ruin monitors. Computer-specific speakers have accounted for that by placing shielding around the magnets, but noncomputer-specific speakers have no shielding. So, if you're using speakers that aren't specifically designed to work with computers, you should place them as far away from your monitors and disk drives as possible. Check the back of the speakers or the owner's manual to find out whether the components are shielded—then place them accordingly.*

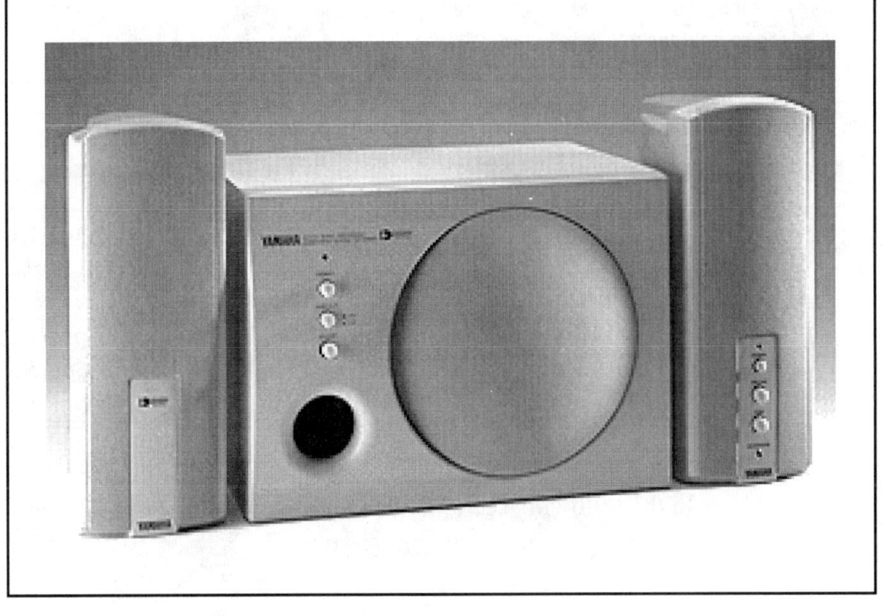

FIGURE 3-1 *The Yamaha System 45 multimedia speaker system*

A Quick Word About Headsets

Some people forego the speaker scene altogether and use headsets or telephone handsets instead. Headsets are quite convenient, but if there's any static on your connection or if the person on the other end has his or her microphone turned way up, your little eardrums will be blasted away. So use a headset only if you'll be talking to people who you know have properly configured their volume levels.

I started out using inexpensive speakers and a handheld microphone that came with my computer, but I found that having to hold the microphone was a real pain and prevented me from doing other things—like cruising the Web, typing a letter, or playing with my Buzz Lightyear doll—while I talked to other people.

Then I purchased a high-quality headset with an attached boom microphone. The first time I tried out this new gadget, the sound was perfect, and the caller on the other end told me my voice quality was superb. But the next time I tried to connect to someone—an architect in Singapore—I almost lost all hearing in my right ear.

As usual, the first thing he said was, "Can you hear me?" Well, not only could I hear him, but the neighbors down the street could too. He had the volume on his microphone turned up to the max, and he was literally yelling into it.

So be careful. I still use the nifty hands-free microphone attached to my headset, but I've left the headset adapter unplugged; I have my external speakers plugged in instead. With this setup, I can hear all my calls very clearly, and my hands are free to do other things.

Sound Cards

A sound card brings life to a mute PC; it is the heart and soul of your computer's sound. Unlike Macintoshes, PCs are not made to have true sound capabilities on the motherboard. They can beep and play sounds (commonly referred to as .wav files in Windows), but they can't process voice or music. So you need to add a sound card to give your computer a voice. A sound card is *the* major factor in determining whether your Internet telephone conversations are crystal clear or utterly inaudible.

Many variables influence just what kind of sound card you should choose. In the following sections, I've highlighted some of the Internet phone-related features to look for as you shop.

Full-Duplex or Half-Duplex?

Have you ever talked on a CB radio? First you speak. Then, when you're finished, you say something like "Over" or "Ten-Four, Good Buddy" and listen for the other person's response. Makes for a S-L-O-W conversation at best.

Or what about those hideous speakerphones? The people on the other end sound like they're in a box, and you have to wait until they finish talking before you can say something, or else you might be cut off mid-sentence.

This clunky, one-person-talking-at-a-time phenomenon is commonly referred to as *half-duplex* communication. It can take a while to get used to, since most of us are used to the *full-duplex* capability the telephone system affords us.

With full-duplex communication, both parties can talk at the same time. You don't have to worry about interrupting the other person. You can still hear what the other person is saying—even if you are speaking too.

In order to get full-duplex flexibility with your computer, you need to have a full-duplex sound card. If you purchased your computer within the last three years, your sound card is most likely a half-duplex model. Check with the card's manufacturer to find out for sure. As recently as February 1996, I purchased a Packard-Bell Force 441CD, and even this computer—a fairly powerful machine at the time—came with a half-duplex Aztech Sound Galaxy card (see Figure 3-2).

for mac users *If you have a Mac Performa or Quadra, you have full-duplex sound capabilities.*

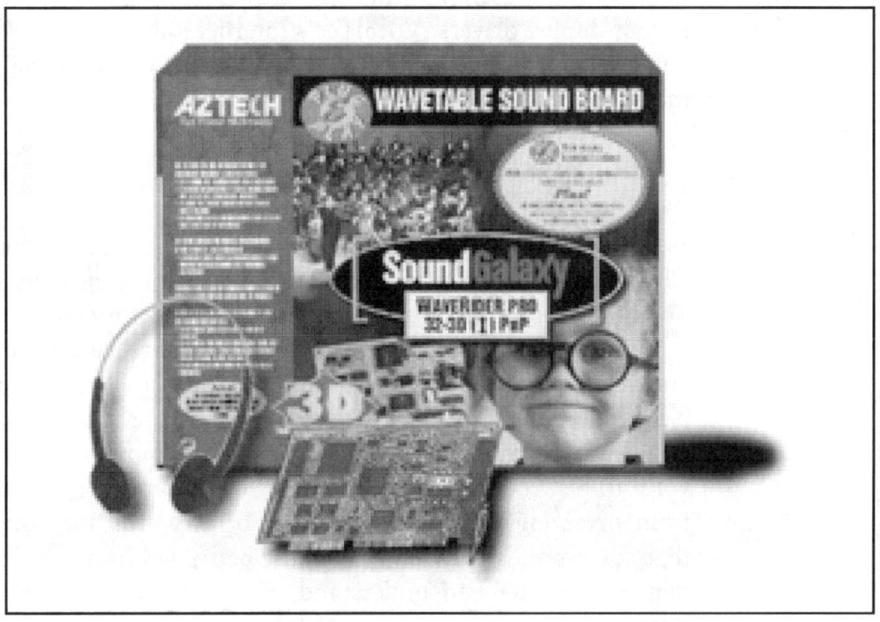

FIGURE 3-2 *The Aztech Sound Galaxy WaveRider Pro sound card*

Today, most high-end systems come with full-duplex cards, but many consumer systems are still sold with half-duplex cards. Before you buy a new system, be sure to ask the salesperson which type of card is included. If he or she is unsure, check the literature that comes with the computer, or check the card's audio settings (through the Multimedia Properties option on the Windows Control Panel) to see if it has two DMA channels. Oftentimes the technical specifications in the owner's manual will also say whether the card is full- or half-duplex.

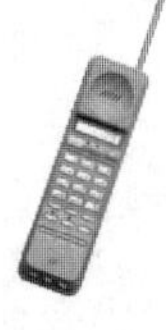

tip *The Sound Blaster 16 and Sound Blaster AWE32 are full-duplex cards that ship with half-duplex drivers.* Drivers, *by the way, are software programs that make hardware work properly with your computer and operating system. The half-duplex sound drivers that ship with the Sound Blaster 16 and Sound Blaster AWE32 can be replaced with full-duplex drivers that take complete advantage of the sound cards' capabilities. You should download these drivers if you want to be able to talk and listen at the same time. You can get these new drivers for either Windows 3.1 or Windows 95 by contacting* ***http://www.creaf.com****. (Be sure you download the right driver for the version of Windows you are using.)*

Most programs, such as FreeTel, Internet Phone, and WebPhone, will show you through their configuration menus whether the program interprets your sound card to be either full- or half-duplex—or, in the case of Creative Labs' Sound Blaster 16 and AWE cards, whether they are utilizing the full- or half-duplex drivers. VocalTec's Internet Phone version 4.0 will even give you a complete rundown of your system; just open the Help menu and run the Support Wizard.

What a Sound Card Does

Your sound card gives you the ability to hear music coming from your CD player, or hear the ringing sound when someone calls you with VocalTec's Internet Phone. But that's not where the story ends.

If you were to create a quantitative chart of the sound coming from your voice box, you would realize that sound can cover a wide range of values. Now, computers are digital devices that can process only ones and zeros (which, respectively, represent On and Off commands). In order for your voice to be processed successfully by your computer, it must be converted from an *analog* to a *digital* signal—in other words, the range of frequencies that corresponds to your voice must be turned into simple zeros and ones that a computer can understand. And after your voice travels over the Internet to another person's computer, it must be converted back from a digital signal into sounds that your friend can hear. This is where your sound card comes into play. It has an *analog-to-digital converter* which turns the voice coming through your microphone into digital signals. These signals can then be compressed by the Internet telephone software and sent through your modem to your waiting buddy on the other end.

This conversion process is commonly referred to as *sampling*. One factor determining how well your voice matches the sampled signal is called the *sampling rate;* this is simply the number of samples made per unit of time. The sampling rate governs the highest and lowest frequencies of sound that can be recorded and reproduced. A low sampling rate provides a less accurate representation of the original sound. The sampling rate is most commonly measured in Hertz (Hz).

tip *Want to see what the sampling rates are for a telephone, CD, or radio? From the Windows 95 Control Panel, select Multimedia Properties. Then click on the customized button for your microphone, and you'll see a drop-down list of different sounds, each with the sampling rate listed.*

Sample size is another factor that dictates how closely your digitally converted voice compares to the real thing. The sample size is the range of values used to represent each sample. It spans the difference in volume between the softest sound and the loudest sound. The sample size is usually expressed in bits, and in keeping with the theory I expressed earlier, bigger is better—the larger the sample size, the closer the sound is going to be to the original.

For example, the standard size of audio quality CD sound is 16 bits; for analog radio it's 8 bits. No one can dispute that the quality coming from your CD player is far crisper and cleaner than that coming from your car radio, where muffled voices are the norm.

tip *You can use Speak Freely's 8-bit sampling option and echo speaker to hear the difference between sample sizes. This is also an excellent program for testing out different sound compression options. You fire it up, connect to an echo server (actually a computer waiting to record and then play back your voice), and in ten seconds the echo server computer plays back the recording of what you've just said. By changing the compression type in the setup of the program, you can hear exactly how close each sample sounds to the original.*

What does all this have to do with buying a sound card? Well, a 16-bit sound card can process more data than an 8-bit card can. An 8-bit sound card's sample size is only 8 bits, so it will reproduce voice less accurately than a 16-bit sound card. To put this comparison in layperson's terms, an 8-bit sound card will sound more like an inexpensive tape recorder or a radio, whereas a 16-bit card produces sound you expect from a CD.

note *Some early models of the widely popular Sound Blaster were 8-bit, but those currently on the market are all 16-bit sound cards.*

Besides looking for 16 bits in your sound card, you'll also want to look for DSP and SNR (or S/N, as some manufacturers call it) in the sound card specifications (which are usually found on the side of the sound card box or within the product manual).

DSP stands for *Digital Signal Processor;* this designation means that your card can perform several tasks at once, like compressing and decompressing sounds. A sound card with multitasking capabilities will take some strain off your computer's processor.

SNR stands for *signal-to-noise ratio.* Every sound card is rated with an SNR decibel level. This level tells you how clean the output of the card actually is. The higher the decibel level, the cleaner the output and the better the sound quality.

Another technical feature to look for that virtually all new sound cards offer is support for PCM, ADPCM, Mu-Law and TrueSpeech. This basically means that there are sound compression features built into the card itself, which allows for faster decompressing of your voice. This results in less delays—the person on the other end will hear your voice more quickly and your conversations will be more like real telephone conversations.

Again, I won't delve into the audio-related components of sound cards, since those features do little to enhance Internet phone communications. If you want to know what some of the techie buzzwords mean, check the glossary at the back of this book.

tip *If your computer isn't equipped with a sound card, you can easily install a good multimedia upgrade kit for about $300. Included in most of these kits are a CD player, a sound card, inexpensive but workable speakers, and the software that makes it all work.*

Portable Options

If you have a portable computer, you can add sound to your mute little laptop buddy. CAT, a little box that connects to your parallel or printer port, offers half-duplex sound quality and a headset with an earpiece and microphone. It also offers two jacks, one for external speakers and another to plug in the headset, for a handheld microphone.

You can try PCMCIA sound cards from Argosy, IBM, Media Vision, or Turtle Beach. These cards slip into your portable and offer true 16-bit sound capabilities. Most of the cards are compatible with any software that works with Creative Labs' Sound Blaster card. Depending upon the package, one of these cards may come with built-in jacks for external speakers and microphones. The average price for these products is $150-$350.

Sound Card Manufacturers

There is a long list of sound card manufacturers, most making some top-quality products. Choosing the right product is more a matter of price than anything else.

The one sound card manufacturer you'll hear about most often in the United States is Creative Labs, makers of Sound Blaster; however, Aztech, Inc., has sold more sound cards worldwide, and their products are included in many consumer systems. If your machine is labeled "Sound Blaster compatible," it simply means that the hardware and software match the way a Sound Blaster card deals with sound. All of the Internet phone programs work with Sound Blaster-compatible cards. You should, however, check with the manufacturer of your phone software to find out exactly what cards are and aren't compatible.

Manufacturers such as Turtle Beach, AdLib, Gravis, and Acer offer high-quality, full-duplex sound cards in the price range of $150-$300. Again, many of the advanced features you might see on these cards—such as Wave-Table RAM, 3-D audio, and MIDI play—have more to do with music and multimedia games than they do with voice, so be sure you're not paying for features you won't have much use for.

Here are some of the companies whose sound cards are worth checking out:

ADVANCED GRAVIS

Phone: 1-800-324-4084
E-mail: sales@gravis.com
Home Page: http://www.gravis.com

Makers of the UltraSound Plug & Play Pro, Advanced Gravis is known for reasonably priced, high-quality sound cards and excellent product support via the Internet. The UltraSound has been tested with a wide variety of Internet telephone products, including VocalTec's Internet Phone, and is on VocalTec's recommended list of full-duplex sound card manufacturers.

CREATIVE LABS

Phone: 1-800-998-1000
Home Page: http://www.creaf.com

Creative Labs markets the very popular Sound Blaster series of cards. The Sound Blaster 16 and 32 both offer crisp audio and relatively easy-to-install software. If you purchase one of these cards, make sure you get the updated driver from the Creative Labs Web site; the SB16 and AWE32 are actually full-duplex cards that ship with half-duplex drivers. (AWE stands for Advanced WavEffects, a term that applies more to music than speech.)

TURTLE BEACH

Phone: 510-624-6200
E-mail: sales@tbeach.com
Home Page: http://www.tbeach.com

Turtle Beach makes an excellent 16-bit, full-duplex sound card called the Tropez that works with all popular Internet telephone programs. It's also a killer sound card to have if you are into music and multimedia games.

Microphones

A good microphone can make the difference between a good conversation and a bad one. Most people you talk to over the Internet use the microphone that came with the computer. Usually these mikes are of fairly low quality and don't offer the clearest sound. Even if you have the top-of-the-line equipment on your end, the voice coming over your high-quality speakers may sound awful simply because the person on the other end is using a cheap microphone or is sitting too close to it. Remember, the weakest link in the chain dictates the quality of your sound.

When you're shopping for a new microphone, you need to get one that matches the capabilities of your sound card. This means you'll need to know a few technical terms. My first recommendation is to check the product specifications of your sound card and see if it requires a particular type of microphone.

Computer sound cards require a fairly strong voltage signal, or signal level—about 10 millivolts, to be exact. Older 8-bit cards need 100 millivolts in order to process the sound coming in from the microphone.

This is why it probably won't do to try connecting just any microphone you find around the house. The standard, noncomputer-type mike puts out about 1 millivolt—not much power at all. If you were to connect this type of microphone to a sound card, you would most likely have to talk really loudly in order to create a strong enough signal for the sound card to process what you were saying. Most sound cards require microphones with an output voltage level in the range of 30-200 millivolts.

Another important factor in buying a microphone is matching the *impedance* of the mike to that of the sound card. Impedance is basically an electrical characteristic much like resistance. The output impedance of your mike must be less than the input impedance of your sound card. Although almost all computer microphones match the impedance range of most sound cards, some professional microphones may not. If you don't have the right impedance, the signal coming from the microphone may not be fully transferred to the sound card. If the impedance of the microphone is too high, some or all of the microphone's strength will be lost, and you'll end up shouting into the microphone. Most sound cards have an input impedance range of between 600 and 2000 ohms.

The two most common types of microphones on the market are *omnidirectional* and *unidirectional*. Omnidirectional does just what the name implies—it picks up sound equally coming from any direction. A unidirectional microphone only picks up sounds that are directly in front of it.

Unidirectional microphones are the preferred type for voice communications over the Net. You'll probably see the terms *cardioid, supercardioid,* and *hypercardioid* on many unidirectional microphones. Cardioid-type microphones all pick up sound through different variations of a heart-shaped pattern of sensitivity (supercardioid and hypercardioid are variations of this heart-shaped pattern). Don't worry—simple cardioid will work fine for Internet telephone connections.

Two more terms you may see that pertain to microphones are *condenser,* (sometimes called *electret*), and *dynamic*. The difference? A condenser microphone is basically battery-powered and is used for recording sensitive sounds, whereas a dynamic microphone doesn't require any external power source and is best for recording loud sounds.

From time to time, I'll use two different microphones, one that came with my Andrea headset (see Figure 3-3), and another cheapie that came with a CD I bought. The handheld microphone, a unidirectional dynamic microphone, works just fine, but does pick up some background noise. On the other hand, the Andrea headset utilizes special anti-noise technology and provides a much crisper, less fuzzy, sound. I use the headset when I know someone is going to call and I want to have a private conversation. But when I'm showing my friends how Internet phone programs work, I use the

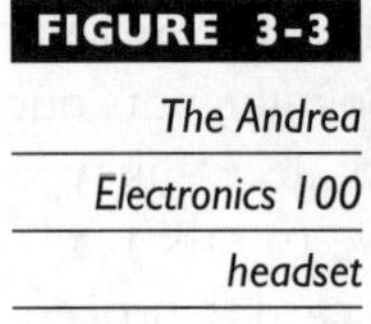

FIGURE 3-3

The Andrea Electronics 100 headset

handheld microphone/speaker combination so my friends can hear what people are saying. If you plan to use Internet telephones in business, I'd recommend using a headset for two reasons: privacy and hands-free talking.

In a Nutshell...

A sound card, a microphone, a fast computer. Beyond the modem, which I discuss in the next chapter, these three pieces are really all the hardware you need to start making long-distance phone calls for free. If you watch what you buy, and don't go overboard, you'll have saved enough money on long-distance charges in one year to cover the cost of your newly supercharged multimedia computer.

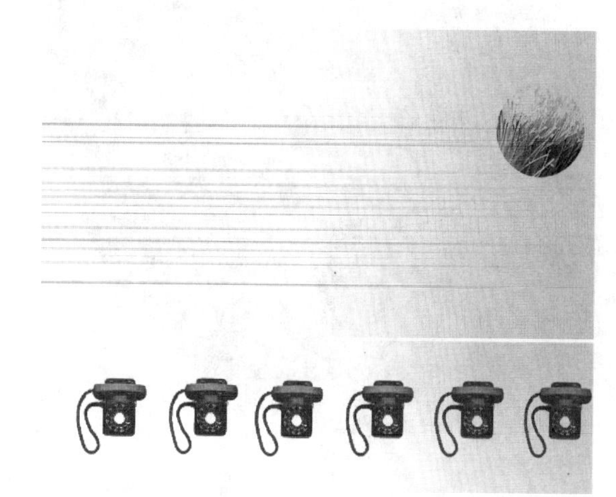

TELEPHONE

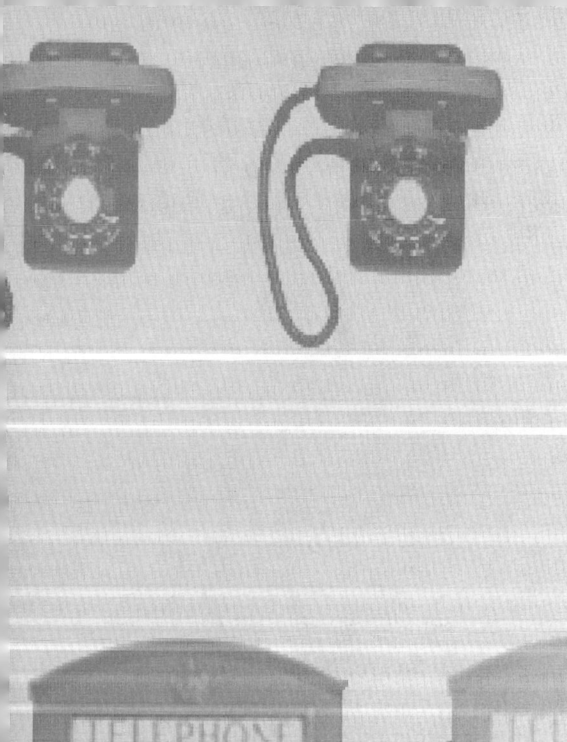

Chapter 4

How to Make the Connection

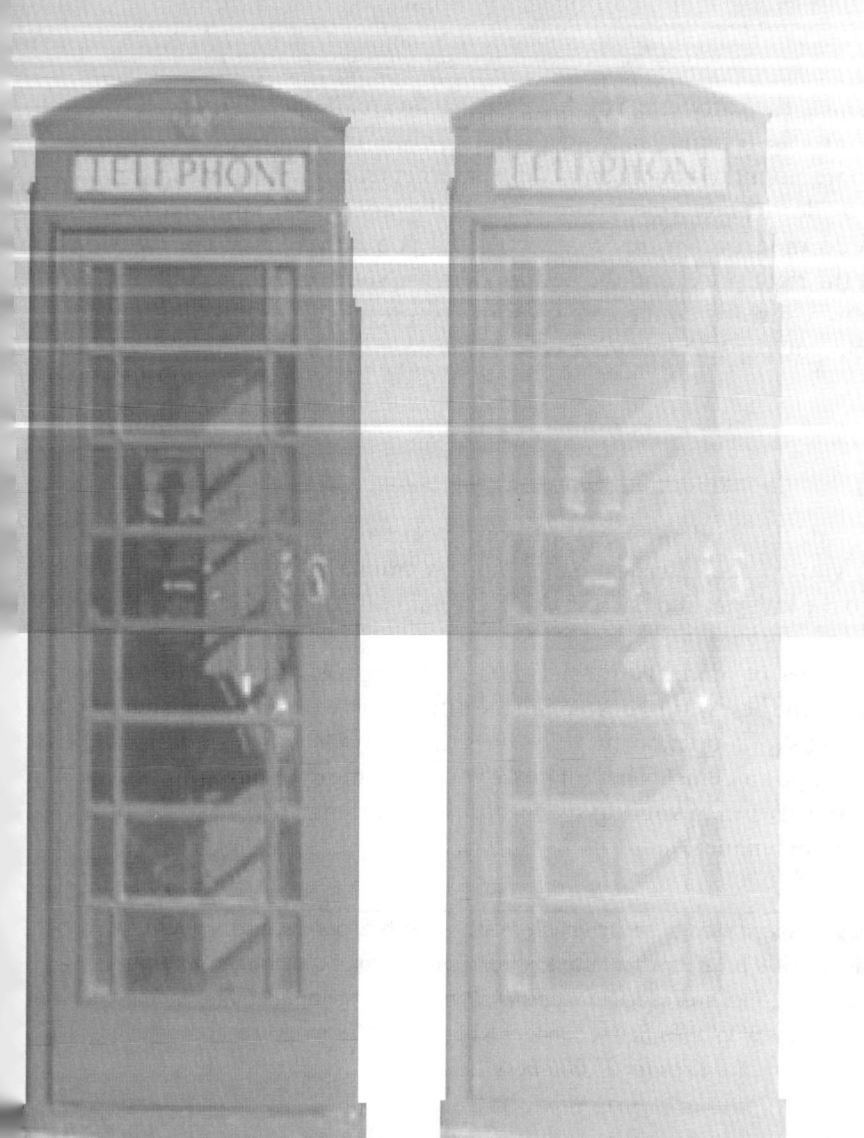

YOU have the software. You have the hardware. Now all you need is a connection to the Internet. Having a reliable, noise-free connection to the Internet is paramount to having an audible Internet telephone conversation. Internet providers, both local and commercial, vary widely in price and quality. So do modems. In this chapter I'll tell you how to find the right provider at the right price, and the best modem for your particular configuration. If you're already connected, I'll show you some nifty tools and tips that will help you supercharge your configuration.

Modems

Let's start with the modem. Think of your modem as an extension cord connecting you to the Internet. Your modem transfers data from your computer to the Internet and back again at a certain rate of speed, commonly referred to as *baud rate*.

Different types of modems transfer data at different speeds. If modems were cars, then the best way of describing the differences would be this: There are Volkswagens (dial-up modems), Porsches (ISDN modems), and Formula 1 Race cars (cable modems). They all move you along the cyberhighway, but at different speeds and in varying styles, and with varying costs in terms of upkeep and maintenance.

note *Modem speed used to be commonly measured in* bits per second, *or* bps. *A modem with a 9600 baud rate, for instance, transferred data at a maximum rate of 9600 bps. Nowadays, with ever speedier modems appearing on the market, we generally discuss speed in terms of* kilobits per second, *or* Kbps. *So a 28.8 modem moves data at a maximum rate of 28.8 Kbps (or 28,000 bps).*

If you already have a 2400 or 9600 baud modem—basically what is considered a bicycle with training wheels—then run, don't walk, to your local computer store and get a faster one. You simply can't operate any of the Internet telephone programs reliably with a modem slower than 14.4 Kbps. Actually, even 14.4 Kbps is almost too slow for the large majority of Internet phone programs. The faster your modem, the faster the data comes back to your computer, which means better sound quality and less delay hearing the other person.

The type of modem connection you select depends on several factors:

- How much you want to spend per month—and in some cases per minute—to be connected to the Internet. (Slow modems can be gas-guzzlers, you know.)
- What types of service are offered in your area. (Sometimes there just isn't a Porsche dealer close by.)
- What types of service are offered by your Internet service provider—the company that connects you to the Internet. (Sometimes the on-ramp is just a one-laner.)
- Whether your computer is powerful enough to really take advantage of the increased speed. (Maybe you just don't have the guts to drive at top speed.)

Let me give you a brief overview of each modem, how it connects, and the pros and cons of each type. Whether you are thinking of getting onto the Net or you've already gotten on and you're wondering what your options are, this section will help you understand how to match your Internet service provider to your unique computer setup.

Dial-up Modems: Cybercars for the Masses

Dial-up modems convert your computer's signals into telephone signals and then transmit those over a standard telephone line. If you've ever heard modems or faxes making those screeching noises, what you've heard is the conversion of digital information into tones the telephone can understand.

Modems work like this: Your computer software instructs the modem to dial the phone. The modem has enough programming built into it to understand those instructions and generate the right tones to dial the phone number of your Internet service provider. It then connects your computer to the service provider's modem and computer, which in turn connect you to the Internet.

When the modem attached to your service provider's computer answers the phone, the two modems check each other out, making sure each can understand what the other is saying. Then they decide on how fast they should communicate, and a connection is established.

note Modem *is actually an abbreviation of the terms* modulate-demodulate.

Dial-up modems come in a variety of speeds—14.4, 28.8, and 33.3 Kbps—and with a variety of built-in compression schemes. Don't even consider buying a 14.4 Kbps modem, and if you have one, consider upgrading. Even though you can use a 14.4Kbps to cruise the Net, the wait will be excruciating and the sound quality very poor. If your service provider only offers 14.4Kbps modems, which some still do, dump them and look for another provider with faster modem lines.

remember *The faster the modem, the more data or sound can be pressed through the phone line. This means fewer delays, fewer dropouts, and a more audible Internet phone conversation.*

The jury still seems to be out on 28.8 versus 33.3. The 33.3 modems are a little more expensive, but do offer more throughput. However, there are two caveats: If your service provider doesn't use 33.3 modems, the additional speed is wasted. And if the quality of your phone line is poor (i.e., if it picks up a lot of interference, or you have bad cabling), it may slow your modem down, regardless of the speed of the modem. The fastest modem can't work optimally if you have cabling or telephone line noise problems. (For example, the wiring in my office is of such poor quality, my 33.3 Kbps modem rarely connects at its top speed.)

My service provider has some 33.3 Kbps modems, but mostly 28.8 modems. It's the luck of the draw whether I get a 33.3 or not, since there is no special phone number I can use to catch my service provider's 33.3 modems. This seems really silly, but many service providers operate like mine. Before you invest in a 33.3, check with your provider and find out if they offer modems with this speed, and if there is a special phone number you can use to dial into them.

Cable Modems: Luxury Cruising for the Lucky Few

At present, dial-up modems are limited by the capabilities of the phone lines. What seemed like a high-speed modem yesterday offers the slowest connection today, mainly because cable companies have started offering Internet service.

Basically, cable modems are little boxes that connect you to the Internet through your cable company, much like your cable box connects you to all those TV channels. Cable modems are the high-speed kings. And just like your TV cable box, a cable connection to the Internet is constantly *on;* in other words, it's constantly available to you. You don't have to call up the cable company and say, "OK, now turn on the Internet for me, I wanna connect," which is effectively what you do go through with dial-up modems.

This also means you don't have to tie up your phone line when you want to use the Net. And you get cable TV on the same wire that gives you Internet data. Cable modem transfer speeds can run up to 256 Kbps transferring data back to you compared to the lowly 33.3 offered by the fastest available dial-up modem.

Your connection to the cable box is made with an Ethernet cable, which means you need an Ethernet adapter in your computer in order to connect to a cable modem. *Ethernet* is a fancy name for a type of cabling system that resembles a fat telephone cord. Ethernet can transfer data up to 100 megabits (that's a lot of bits!) per second. It's the same kind of cabling technology your office probably uses.

Most personal computers, particularly home PCs, don't come with Ethernet connections, so you may need an adapter card if you plan on using a cable modem. Ethernet adapter cards usually run between $100–$250. Ethernet adapter cards attach to your computer's main circuit board, which means you'll most likely have to remove the cover of the computer to pop in the board. If you are buying a computer for the first time and are thinking of using a cable modem, make sure the technician installs the card first and loads the appropriate software to make it work.

for mac users *Most newer Macintoshes, including the PowerMac and Quadra computers, include built-in Ethernet connections. This means that if you have a Mac, all you need is a cable to connect to a cable modem.*

Since your cable company is providing the connection, they are also providing your Internet service, which normally includes e-mail, Web, and file transfer capabilities. Presently, and unfortunately for those of us who would love to have a high-speed connection, few cable companies offer cable modem connections. This is due, in part, to the fact many cable companies must upgrade their equipment—and in some cases, cabling—to accommodate two-way communications. But as equipment, wiring, and technology are upgraded, the cable companies will be in the best position to offer Internet users the fastest type of Internet connection. Within the next few years you should see more cable companies offering this service. And if your local cable company offers cable modem connections, I highly recommend you sign up for the service. You won't be disappointed. One caveat though: If you want to connect to another Internet phone user or simply want faster access to the Web, understand that all the speed in the world on your side doesn't do much if the connection on the other end is slow.

ISDN Modems: Middle-Class Cruising

ISDN stands for *Integrated Services Digital Network,* which is just a fancy way of saying combined digital telephone services. ISDN combines your phone, fax, and Internet connection into a single phone line—the same copper wire that now comes to your house to deliver your telephone service.

ISDN also provides true digital telephony, and that means there is no need to translate computer signals into telephone signals, meaning faster data transfer rates. Your computer's digital signals go straight from your computer to the telephone company's switch, and then out to your Internet service provider. Like cable modems, it's like having a long cable connecting you to the Net. The result is a far faster, far clearer, connection than you could ever achieve with standard analog modems.

ISDN connections can run at up to 128 Kbps. To obtain this kind of speed, you'll need an ISDN modem or ISDN adapter, and an ISDN-capable telephone connection. If you live in a relatively populated area, your local phone company probably offers ISDN by now. But if you live in a remote area, don't count on ISDN becoming available any time soon. Your telephone company will first have to upgrade their telephone switching equipment, and without enough potential users to offset the cost, most rural phone companies can't afford to offer ISDN.

ISDN also costs way more than either cable modems or standard telephone lines. Most phone companies charge for every minute you use your ISDN connection during peak hours. In addition, installation of ISDN connections can run up to $500 per line, and ISDN modems run about $250.

If you live very far from the phone company's equipment expect more charges to be tacked on as well, or service not to be available at all.

An ISDN modem either connects to your serial port or can be installed internally as an adapter card, much like an Ethernet adapter card. As with a 33.3 Kbps modem, in order for you to utilize your ISDN connection, your Internet service provider must also have an ISDN modem on its end. Most service providers charge more for using this type of higher-speed connection, since they have to invest in additional equipment. Plus, the more high-speed users your provider has, the fewer low-speed users they can accommodate, meaning that they'll have to add more connections from the Internet to their system. This all means added costs.

remember *Whatever connection you use, if the person you are talking to doesn't have the same type of connection, all that speed is wasted. If your daughter connects to the Internet with a 28.8 Kbps modem and you have an ISDN line, your conversation will be dragged down by her slow connection.*

A Few Words About Service Providers

Over 60 percent of all the people connected to the Internet use local or commercial service providers via their home or businčss dcsktop computers. You'd think the simple act of connecting your computer to the Internet would be relatively easy and that service providers wouldn't vary much, but not all Internet service providers are created equal.

From a consumer standpoint the only difference you may see up front is price, but from a technical standpoint the differences can be tremendous.

Typically, prices range from $15 to $30 per month. But beware! Some providers, especially commercial ones, tack on per-hour charges as well. Check up front to find out the exact costs, not only for the monthly fee but for hourly fees or network charges. For example, America Online touts 10 free hours each month. But in Alaska, although the call to connect to AOL is local, America Online assesses a network charge of six cents per minute. So for us Alaskan users, "free" really means $3.60 per hour.

If you don't spend much time on the Net, you may want to opt for an hourly payment plan. But if you are like many surfers, hopelessly hooked, you'll want a flat per-month rate—one that doesn't change regardless of the number of hours you may spend on the Net. Using the Net can become extremely addictive, and costly if you use commercial services.

Knowing how to evaluate and choose between different providers means more than just knowing which payment plan will work for you. You'll also need to know a few technical terms, and understand how a service provider connects you to the Internet.

Choosing a Provider That's Right for You

Your first decision involves choosing between the three different types of providers: local service providers, national service providers (such as Netcom), or national commercial services (such as America Online, CompuServe, and Prodigy). Local providers usually offer competitive pricing, but don't offer nationwide access. Local and national providers usually provide software to get you onto the Internet, but they don't provide much else. Commercial services, however, offer deluxe features such as chat sections, specialized news services, and graphical interfaces to simplify cruising the Web. Commercial services are like cable TV.

But commercial services have a downside to them. Unlike both local and national service providers, commercial services oftentimes don't provide you with true Internet connections, and require that you use their own specialized software that connects you to their system. A true Internet connection turns your computer into an actual node, or addressable computer, on the big Internet network. With a commercial service, your modem dials into their computer, meaning your computer essentially acts as a dumb terminal, not a networked computer. Internet phone programs need to be able to communicate directly from computer to computer. If you use a commercial service provider, your computer doesn't appear as an addressable computer and you may not be able to use many Internet phone programs.

Commercial services' software doesn't use standard Internet commands, and this can cause problems for Internet phone programs that are looking for true Internet connections. Intel flat-out states that their Intel Internet Phone product will not work with providers such as America Online, since AOL does not provide users with true Internet Protocol (IP) addresses.

That's not to say you can't still use many other Internet-type programs. But you will have a harder time of it, and you'll probably need to download special software that will cause your commercial service to work with your Internet phone programs. In addition, having another computer between you and the person you are calling can slow things down since the commercial service provider computer has to process your instructions and pass it along to the caller.

If you must use a national or commercial service because you travel, or because you want to use the special forums they offer, I prefer CompuServe over all the other commercial service providers. CompuServe offers easily configured, true Internet connections through their PPP dial-up service. However, you will find CompuServe to be much slower in response time compared to a local Internet service provider. If you don't travel much, and have no plans to connect on the road, I suggest you seek out a local provider. Their connections are oftentimes much faster and usually much cheaper for unlimited connect time. If you do find yourself traveling a great deal, and aren't interested in special forums that commercial services provide but still want to use your Internet phone software to call your loved ones and business associates, you should consider a national provider. Since national providers offer either 800 numbers or a wide list of dial-in access numbers throughout the country at about the same price as a local provider, they are the way to go for the traveling Netphone fan.

What Is This "Dynamic PPP Connection" All These Service Providers Are Talking About?

Every computer on the Internet must be assigned a unique number so that data can be routed to the proper place. This number is called an Internet Protocol address. *A* static *IP address is one assigned to you and you alone. A* dynamic *IP address is one that changes every time you log on. With dynamic IP addresses, your service provider has a pool of addresses, and hands them out on a first-come, first-served basis each time somebody logs on.*

You really don't have to worry much about IP addresses. Most software today figures out your IP address and sends information to the right place. But if the software doesn't have a directory service to help connect users together, much like CU-SeeMe, it can be difficult to connect with the other person if your IP address changes each time you log in. You literally have to e-mail your IP address so the other person can call you, since the software itself wouldn't do that. This isn't a problem, however, if the other person you are calling is using a static IP address. But again, very few people are connecting to the Internet via static IP addresses.

The best analogy I can think of is a regular telephone. A static IP address is like your home phone number: it never changes, and people always know where to reach you. A dynamic IP address, however, is like you using different pay phones each time you want someone to call you. In order for people to contact you, you have to continually tell them what phone number you're at and how long you plan to stay there waiting for their calls. Today, many programs compensate for dynamic IP address users by providing "directory services" or listings that show the person's name, while behind the scenes keeping track of the current IP address you are using. This allows people to call you without having to know anything more than your name.

Whether you get a static IP address or just use the most common dynamic type should be based on two things: whether the Internet phone program you are using only works with IP numbers, like Internet Call or PGPfone, and whether you plan to use that program frequently to talk to other people.

You really don't need a static IP address, since most of the Internet phone programs provide some form of a directory service that lists all the program's current users. Consider too that a static IP address costs a lot more, so unless you really want to use a particular program and want to be trendy and hand out your IP address like you do your phone number, then by all means, sign up for that cheaper dynamic one.

And what specifically should you look for in an Internet service provider? They seem to be popping up all over the place, and making the right choice means making sure they adhere to some basic requirements.

Requirement 1: Customer Service

Besides price, the first thing to look for is good customer service. A provider can have the most high-tech equipment available, but if the support staff can't tell you how to configure your MacPPP connection, then none of that machinery will help you one bit.

If a provider doesn't answer the phone, puts you on hold, or can't answer your question in common English, you may have trouble down the road. Of course you may not ultimately experience many problems, or have any need for your provider's technical support once you are connected, but it's getting connected in the first place that can be frustrating.

Requirement 2: Compatibility with Your Computer

Check to see if the provider you're considering offers preconfigured connection software for the type of machine you are using. My Dad's first service provider didn't know a thing about Macintoshes, so they were no help at all when it came to installing the Internet connection software on Dad's Mac IIci. It took Dad several telephone calls to his own personal technical support line (me) before he was cruising the cyberhighways. And this costs him money. (I charge LOTS for tech support, especially if the customer is a relative.)

If your prospective provider doesn't cater to the type of machine you have, it will be more difficult for you when you decide to upgrade equipment or when new connection software comes out for your particular machine. A good service provider will cater to both PC and Mac platforms and maintain a file server with all the latest upgrades to browsers and connection software for both machines.

Requirement 3: A Powerful Connection to the Internet

The Internet is a whole bunch of computers connected together. Your computer is connected to your service provider's computer, and their computer is connected to yet another network provider. The bigger the "pipe," (bandwidth) your provider uses to connect to their network provider, the faster the information will flow from the Internet to your service provider back to you.

Let me give you an example of three different service providers in my area. Service provider A is a the largest. They have one T3 45-megabit-per-second link connecting them to the Internet, which makes them roughly the equivalent of a six-lane highway with a speed limit of, say, around 4000 miles per hour. (Hey, the information infobahn!)

Service provider B connects to the Internet through a single T1 1.544-megabit-per-second line—basically a four-lane highway with a 135-mile-an-hour speed limit.

Service provider C actually connects to service provider B, through a 56K connection—effectively a two-lane highway with a five-mile-per-hour imposed speed limit.

If service provider C has a lot of customers, traffic jams will occur, and your cruising will be stalled to a slow crawl. The five-mile-per-hour speed limit indicates the maximum speed you'd be cruising if no one were on the highway to begin with, so cut that down based upon the number of concurrent users on the system at any one time.

But much the same can be said for service provider A. At last count they had around 5000 subscribers, compared to less than 20 for service provider C. If several thousand people get on A's highway, traffic jams are just as likely to occur, and what seemed like a fast connection can actually seem as fast as molasses and response times can be as slow as provider C. On provider A, even though everybody can travel twice as fast as C, if hundreds of people are on the highway, the speed limit doesn't become much of a factor.

So which type of provider is the best one for you? I'd say, first ask people who might be using A if they get a lot of busy signals or if the system seems to slow down during peak hours. If the provider doesn't keep adding capabilities to match the number of users they are accommodating, then even the biggest pipe can seem like a drainpipe, with everyone clamoring to get on.

Provider B may be your best choice if the number of users more realistically matches the size of their connection to the Internet.

Avoid Provider C altogether. They simply don't have enough bandwidth to accommodate you or the heavy traffic you plan to pump through their pipes.

And if none of them can speak to you in English and instead try to impress and confuse you with technospeak, ditch them all and look for one that can, even if it's a national provider.

Requirement 4: Powerful Equipment

The equipment and number of servers your provider uses also dictates how fast your connection will work for you. Even if your provider has the biggest pipe available, if they are running their server off a 486/33MHz PC with four megabytes of RAM, there is no way the machine can physically keep up with the number of requests it has to process.

In the server business, most serious providers use a variety of Sun SparcStations, Pentium-based Windows NT servers, PowerMacintosh servers, or UNIX-compatible servers. The main thing you want to look for—and then steer clear of—is a service provider using a desktop-type configuration, with limited RAM (under 32 megabytes).

In my locale, service provider C actually uses a Macintosh Quadra with a IIsi as a backup. These are very slow desktop machines that can't provide enough horsepower for your voice-intensive needs. To make matters worse, their server works double duty, functioning as a Web server and an e-mail server. When you are ready to press that "TALK" button and start talking, your voice may be delayed because someone else on the system is downloading a huge file from the e-mail server. The best way to run a system is to have a separate server for each function instead of trying to overload systems. So if provider B has a separate mail server from their Web server, then they would easily be the more powerful of the two.

Requirement 5: Reliability

As I write this I'm sitting here waiting for my service provider, the biggest of the bunch, to come back online. The Web and mail servers have died again for the third straight time this month. I can't tell you how many times the connection to the lower forty-eight has been cut, allowing me to connect to my Web server, but preventing me from checking out other servers around the globe or letting me chat on one of my Internet phone programs. The phone lines that get me into the system are constantly busy.

The last time my provider's mail server went down, I lost *a lot* of e-mail—e-mail from people I may never know sent me e-mail. In addition, I couldn't call my folks during our arranged time, nor could I send them an e-mail message to tell them of the technical problems I was having. So I was forced to use a real telephone to make that long-distance call, something I hadn't done in quite some time.

My provider may have the biggest pipe compared to all the others, but it doesn't mean diddly-squat to me when it goes down. As their subscriber base increases, it appears as if the problems are increasing as well. Although they've added technical staff, the depth of total knowledge may be shallow, and many of the staff are relatively young and inexperienced at giving good quality technical support. They also lack a certain professionalism despite understanding that I don't use the Internet for just fun and games. I use it for business communications and it's critical that the system is reliable.

Reliability and performance can almost always be tied to the technical expertise of those who are running your service provider, and smaller, homegrown companies can't always pay top dollar to attract the best people.

Requirement 6: A Direct Connection to Your Machine

This is something that is often overlooked and never really mentioned, but it can be the difference between a clear Internet phone connection and a choppy one: The closer your provider is to the Internet and to the other person you are calling, the more likely it is that your voice will come through crystal-clear.

The distance between connections is referred to as a *hop*. Let me give you another real-world example. My Dad started out with one provider, the first one to offer local service in his area. The cost was $30 a month, and for a while it seemed the service was pretty good. Then more people joined and it became nearly impossible for Dad to log on. In addition, the system he was connecting to, much like mine, started to go down at the most inopportune times.

Finally Dad decided to try another provider, which by that time was offering a $20-a-month service with all the same features: e-mail, Web browsing, file transfer, and a place for his home pages.

This service seems to be as fast if not faster—and in reality, it is. Even though the two providers are located in the same town, just a few blocks from each other, Dad's old provider actually routes through four more network connections than his current provider. Definitely shop around. When Dad first got on there was no other option, but now that the Internet has become a household word, there is plenty of competition to choose from.

Here's one other thing to consider. Since so much competition has sprung up lately in the service provider business, and since many of these providers are more technically oriented than business oriented, you'll see many of the small to midsize services fold. Not because of any technical problems, but because many simply don't have a good grasp on the business end of the business. Many providers are more interested in hooking up a new server than collecting past-due payments. And as we all know, that type of neglect can lead only lead to problems.

Tips and Advice

Once you have picked a stable Internet provider, you should have no problem getting on the Internet or connecting to other Internet phone users most of the time. But even with the best provider there will be those occasions when you might be able to connect to the Internet, but can't make a good connection to another Internet phone user. Perhaps your conversation will sound something like...

M ine. Ow r u? S e eather? How ose ubs?

when it should be...

I'm fine. How are you? How's the weather? How about those Cubs?

It's times like these you need special software tools and tricks to figure out if the problem is coming from your end, the other person's connection, your service provider's server, the Internet Phone directory server, or the Internet. Here are a couple of tools you can use to get a better picture of what is going on, and some remedies to fix those problems.

But remember, these are just the simple tools that give you a glimpse of what's going on behind the scenes. Many factors can affect an Internet connection, and even though these tools can show you how your connection travels from one location to the next, the statistics you get from these programs can change from one Internet phone call to the next and from one Internet connection to the next, mainly due in part to the nature of the Internet. Just remember when you use these test programs understand they are meant mainly to give you just an insight or starting point to help pinpoint possible problems.

Before You Start Testing Your Connection...

What I recommend you do first if you seem to have a bad connection is to disconnect from your service provider. Then turn off your modem, turn it back on, and try to reconnect to your provider again. Sometimes your modem needs resetting, and sometimes your computer does. And sometimes it's the luck of the draw; you may be dialing into a bad modem on your Internet service provider's side, or you may be getting a noisy phone line. The only thing you can do then is try to reconnect and hope you get a different modem or a clearer line. If you continue to experience problems, you should notify the technical support department for your ISP and let them know you've tried several times, even resetting your modem before attempting each connection.

Use the Traceroute Command to Check on a Server

When you connect to another Internet phone user, the connection isn't from point A to point B, but rather from your computer to your ISP's server, then from a router to another router, maybe to another computer, then to another router, and so on. On the Internet you never really connect directly. To see how many connections it takes to get from point A to point B, you can use the *Traceroute* command. This is a network command that checks to see how many hops it takes to get from one point on the Internet to another. For example, a message sent from my provider to Dad's old provider took 13 hops, or jumps between different network devices, to reach Dad. With his new provider it only takes 10 hops between his computer and mine. That means three less places where things can go wrong, or where my voice transmission can be slowed.

tip *You can run the Traceroute command from a DOS window with Windows 95. You would type* ***traceroute domain****.*

for mac users *You can also traceroute with Macs, provided you have either Mac Traceroute or What Route. Both are shareware programs that can be found at* ***http://www.shareware.com****. Just search for all Macintosh software using the keyword "Traceroute." The latest versions of these programs should be displayed.*

When I want to check how many hops it took to get from my provider to Dad's I would type **traceroute flex.net.**

I then get something like this on my screen:

```
1  ict-fuhsd-sccoe-T1.ict.org (206.197.121.1)  1.309 ms  1.26 ms  1.169 ms
2  NBIIa-fuhsd.sccoe.k12.ca.us (204.88.159.13)  19.328 ms  12.898 ms  14.067 ms
3  sl-ana-3-S3/2-T1.sprintlink.net (144.228.73.97)  29.229 ms  29.981 ms  37.79 ms
4  sl-ana-2-F0/0.sprintlink.net (144.228.70.2)  58.361 ms  30.285 ms  261.279 ms
5  sl-stk-6-H2/0-T3.sprintlink.net (144.228.10.25)  37.435 ms  37.717 ms  42.627 ms
6  sl-stk-nap-H2/0-T3.sprintlink.net (144.228.10.50)  40.59 ms  52.398 ms  42.541 ms
7  pb-F1.MCI.net (198.32.128.197)  49.309 ms  118.845 ms  41.48 ms
8  core3-hssi3-0.SanFrancisco.mci.net (204.70.1.201)  43.229 ms  46.73 ms  43.47 ms
9  core3.Dallas.mci.net (204.70.4.13)  121.054 ms  290.189 ms  255.049 ms
10  core1-hssi-3.Houston.mci.net (204.70.1.122)  117.955 ms  126.283 ms *
11  border4-fddi-0.Houston.mci.net (204.70.3.99)  99.062 ms  104.174 ms  94.949 ms
12  flexnet-inc.Houston.mci.net (204.70.39.50)  123.197 ms *  133.832 ms
13  * tron.flex.net (205.218.188.1)  105.536 ms *
```

In each line, the first column simply keeps track of the number of hops it takes to get from point A to point B. The second tells me the assigned name of the machine, if there's a name available. The third number, which is divided into four sets separated by periods, tells me the actual IP address assigned to the machine or router. The next column tells me, in milliseconds, the time it takes for each of three trace packets to reach the specific gateway and return the result.

note *The numbers generated by Traceroute may vary widely as different network functions take place and the overall network load increases and decreases.*

Any lost packets are indicated by an asterisk (*). There are all sorts of reasons for lost packets. Some gateways or machines may not return the appropriate message requested by traceroute, thus making them seem unresponsive, while firewalls, or devices that prevent users from gaining access to company networks, may use packet filters that block the type of packets used by traceroute. If you happen to be behind a firewall that does block traceroute packets, the results show the route to your firewall, followed by a line of asterisks. You do not see the hops it takes outside your firewall. Finally, packets just may be lost as a result of heavy network traffic, a common occurrence lately.

Pinging and Other Fun Things

Another way to check out local competing service providers is by pinging them. It's kind of like submarines pinging to see if there is another submarine in the area.

Ping is a program that sends a request to another computer and waits for that computer to reply, saying that it's reachable across the Internet. With the Ping program, you can find out if a specific computer is up and running. This can help you locate problems when you are using a direct-connection program, such as PGPfone or Softfone, to connect to a particular IP address. If your friend gives you 204.17.155.47 as his computer's IP address, but the program never connects, you can ping that address to find out if it is up and running.

When you ping a domain—not an individual machine, but a group of machines—like servcom.com, you are sending out one byte of information that, if the service is running, will be bounced back to you. It's just like the process that submarines use to find other submarines.

Although Ping is a UNIX command, you can use the following nifty little Web page, written by Nelgin Reed, to ping a site, or even *finger* a user (which means to check and see whether he or she is logged on). This particular Website is located at the Institute of Computer Technology, whose domain, or network designation is ict.org:

http://www.ict.org/~nelgin/utilities.html

This site lets you ping from their server to whatever server you type in. That way you can compare two geographically located sites to see how long it takes to get from the ict.org to the sites you are pinging.

Or you can check my home page, **http://www.netphones.com**, for a list of other Ping-related sites. Nelgin's, along with the other Web-based Ping sites, issue the Ping command from that particular server to the IP address you specify. Ping sites are great if you aren't having problems getting to various Web servers, but they may not work if the problem lies within your own network or your own service provider.

If the problem is within your own network site and you can't get to Nelgin's page, you can issue the Ping command from a wide variety of Windows or Macintosh software designed to let you ping directly from your computer to another specific computer or domain.

MacPing and the Ping utility provided with Windows 95 are two good examples of individual "client" ping programs. With the Windows 95 Ping program you can run it from a DOS window by simply typing PING and the IP address you want to ping.

Let me give you an example of what I receive when I can get to Nelgin's page and want to Ping my local service provider. When I select the Ping option from the Web page, this is the result:

```
PING alaska.net (206.149.65.3): 56 data bytes
64 bytes from 206.149.65.3: icmp_seq=0 ttl=242 time=240.1 ms
64 bytes from 206.149.65.3: icmp_seq=1 ttl=242 time=304.8 ms
64 bytes from 206.149.65.3: icmp_seq=2 ttl=242 time=236.4 ms
64 bytes from 206.149.65.3: icmp_seq=3 ttl=242 time=350.8 ms
64 bytes from 206.149.65.3: icmp_seq=4 ttl=242 time=231.1 ms
--- alaska.net ping statistics ---
5 packets transmitted, 5 packets received, 0% packet loss
round-trip min/avg/max = 231.1/272.6/350.8 ms
```

Notice how long it takes to echo back my one byte of data? Those figures can change drastically depending upon what is going on at Nelgin's site, ict.org, or what processes alaska.net has running at the time, and what's going on at all parts in between. Although Nelgin's page gives you some basic information, you'll find that some of the specific Macintosh and Windows utilities provide you with additional information and the ability to specify different parameters. If you have the interest, I'd recommend trying out various personal computer-based ping programs. Again, check my home page for a list of ping utilities for both platforms, or search **http://www.shareware.com.**

If you'd rather leave the real techie information to someone else, but still find Web page-based network monitoring tools easy enough to handle, another nifty page that will monitor your site and report average packet or information loss is Onlive.com's "Test your connection to Utopia's" page:

http://www.onlive.com/cgi-bin/nettest.cgi

When you bring up this page, you'll notice that its sole purpose is to test your connection to another computer, Utopia. If you want accurate results make sure you aren't checking e-mail, downloading files, or doing anything but letting this page do its magic, as shown in Figure 4-1.

The page will automatically reload, checking the time it takes for your connection to communicate with Utopia and recording those results within the Web page itself.

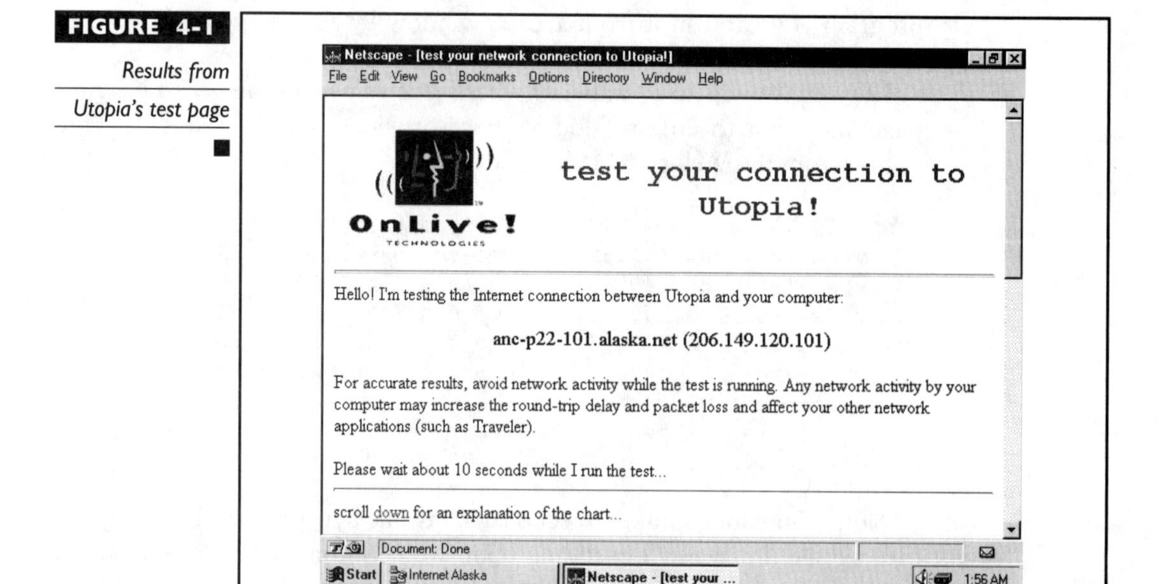

FIGURE 4-1

Results from Utopia's test page

By monitoring packet loss and reporting to you the average amount lost and the time it takes to send packets back and forth, Utopia can help you judge whether or not your Internet provider is having a bad hair day.

What packet loss means to you is choppy conversations, long delays, and slow connect times to directory services that provide lists of other Internet phone-connected users. More than a 30 percent loss, and you probably won't have much luck having clear conversations. With less than a 10 percent loss you should have fairly clear, audible chats with just about anybody anywhere, provided their service isn't experiencing problems.

So Remember...

The key to a good Internet phone conversation is a good Internet connection. There are clearly many variables to consider when deciding on a provider. If you can find one that offers cable modem access for relatively cheap prices (around $40–$60 per month for unlimited access), go for it. But if you are left to your own dial-up devices, check around. Have your Net-connected friends run comparisons between local competing providers. Then make your decision based on speed, customer service, and *last of all,* price.

TELEPHONE

Chapter 5

A Few Words About Sound Compression

YOU really don't have to know much about sound compression to get the majority of the Internet phone programs to work. But knowing a little bit about what's going on with your voice in the background while you blab on and on can't hurt. In this chapter I'll delve deeper into how sound travels from your computer, through the Internet, to the party you are calling. I'll outline the various sound compression algorithms that make all this chatting possible. Plus, I'll highlight how some programs let you fiddle with sound compression to get almost crystal-clear sound. And I'll try to do all that without a lot of technospeak.

It May Sound the Same but It's Not

Your Internet phone program may act like a phone, sound like a phone, and even look like a phone on your screen. But it doesn't transfer sound like the telephone network does. Although both the Internet and telephone networks can handle voice, they do so very differently.

Analog vs. Packet-Switched Sound

Because of advancements in compression, the sound quality coming from today's Internet phone programs rivals that of the telephone. And that's no small feat when you consider that the Internet is constructed and works far differently than the telephone network. The long-distance telephone network is based on analog signal switching. What does this mean in layperson's terms? Simply that the telephone network provides a continuous stream of data to, or a connection to, the other party so you can have real-time, two-way conversations. Analog signal switching takes your voice at face value and simply passes it along down one end of the phone to the other. All the telephone network essentially does is complete that link between you and the party you are calling.

The Internet, on the other hand, is a digital, packet-switched network. This means that your computer first has to translate your voice into digital signals, then wrap those signals into little packages so they can travel on the digital highway. Sound isn't continuously traveling back and forth from computer to computer, as it does from telephone to telephone:

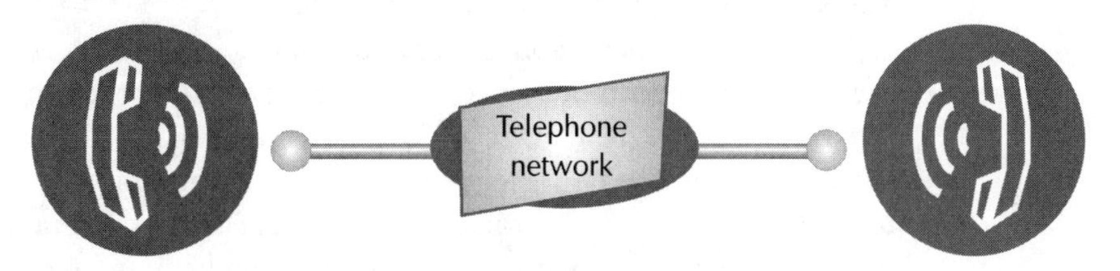

On top of that, when you use the Internet to transmit voice, your little sound package has to travel through a maze of different computers, routers, and servers before it reaches its final destination.

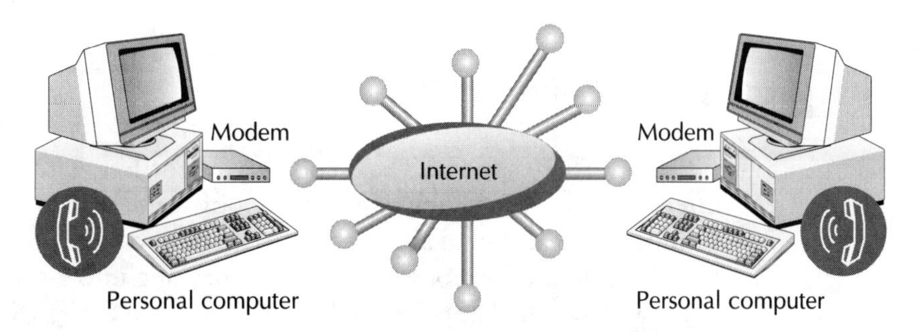

In a traditional telephone network there are also various hardware components that your voice travels through, but their sole purpose is to act as switches—that is, to transmit sound. This contrasts with the usual purpose of the menagerie of machines that make up the Internet. The real function of computer servers and data routers on the Internet is to transmit data and ensure that it gets to its final destination. Computers and routers on the Internet are really meant to transfer bits and bytes, not continuous streaming sound.

So when you use your computer to talk into an Internet phone program, the sound you make is turned into a sound file, and then the file containing your voice is passed from your computer to the next one along the chain. There may be several computers it has to pass before it gets to the other party you are talking to. Also, each computer it encounters along the way may be tasked with doing other things at the time, such as passing other types of data along to another computer or group of computers. This means that your voice file may be delayed getting to the other computer. The file that makes up your voice packages oftentimes gets waylaid. It may be that Internet transmission has slowed to a crawl because of heavy usage. In some cases, your voice may get lost altogether. Servers and routers can go off-line, causing dreaded *dropouts,* or unwanted gaps in the conversation that can last anywhere from a fraction of a second to several seconds. Anyone who has ever called someone long-distance knows that it's very rare and very noticeable to hear long pauses or have parts of a conversation drop out entirely. This is because with conventional phones, the two calling parties are basically connected directly together.

But with Internet phones, dropouts and lost parts of your conversation are somewhat common, since you are connected to each other not directly, but rather through a maze of computers. And herein lies the difference between talking on the telephone network and talking on the Internet telephone network. The former was designed to continuously feed sound back and forth without much interest in error checking, whereas the Internet was designed for data transmissions that require error checking. With all these potential bottlenecks you can quickly see how difficult real-time, two-way conversation is to achieve.

If the Voices Sound a Little Weird...

But wait! There's more! That modem you got free with your Ginsu knife offer is also a major culprit in slowing down transmission of your voice—turning the clear digital audio your computer is capable of producing into a sound much like someone makes when speaking into a trash can.

How does this happen? Well, even if you have a fairly fast 28.8 Kbps modem with compression technology built in, it can still only send 3 kilobytes per second, and that's on a good day. True telephone-quality sound requires about 8 kilobytes per second. That means your modem would have to pump almost three times as many kilobytes of data per second as it currently does to match the transfer rate of a conventional telephone.

The superior quality of the standard telephone system today is achieved by a combination of continuous streaming sound and high-speed transfer rates. In order for Internet telephones to even come close to telephone-quality sound, they too have to compress the sound, and do it more effectively and efficiently than conventional phones. Plus, they need to do it faster so that the sound file can quickly reach the other party without much delay.

The Magic of Data Compression

Without audio compression we Internet phone users would all sound like Charlie Brown's teacher: "Wonk wonk wonk wonk wonk." So how does compression really work?

No matter which product you use, every Internet phone program uses some sort of compression software scheme. Each product must have a compression scheme so it can record your voice, turn it into digital signals, compress those signals into smaller packets of data, and then send those packets to the other computer you're calling. Then those packets are decompressed, turned back into analog sounds, and played back on the recipient's computer speakers. Whew!

In order to make your conversation sound as much like a telephone conversation as possible, the compression program itself must minimize the delay involved in compressing and decompressing the sound. Different compression algorithms, or computations, vary in the amount of compressing they do, how well they do it, and how fast they do it. A compression program that produces high-quality sound requires a fast computer. One that produces low-quality sound, obviously requires a less high-powered machine. The faster the machine, the better the compression software can perform its function with little or no delay.

tip *If you are having problems understanding all this computer babble, close your eyes and imagine you're standing in a line where people are handing you sandbags to pile onto a flatbed truck. The more muscle you have the faster you can heft those bags onto the truck without making the person behind you wait. The wimpier you are, the fewer sandbags you can stack on the truck. So it goes for sound compression programs and computers. The more muscular, or high-powered, your computer is, the faster it can pass along the data to your speakers—and from your lips to the caller on the other end. A hefty computer, such as a Pentium or a PowerMac, can compress outgoing voice packets and expand incoming packets more quickly than less robust machines, thus cutting down on pauses and delays.*

Let's back up a minute and talk about how the Internet transfers ordinary data, like e-mail or computer files. When you transfer a file from one computer to another with a file transfer program such as Fetch or WS_FTP, the sending computer then checks with the receiving computer to make sure the file has arrived intact. But Internet phone programs don't go back to check—or what is called request packets (think sandbags)—to find out if any have been lost (dropped by you or whoever was handing them to you back on that flatbed truck). Instead, phone programs try to wrap up all the sound in a complete package, then send that package along its way, hoping that nothing gets lost. The slower this happens, the more broken your speech will sound to the person you're talking to. The faster the computer, and the faster the modem, the less likely it is that your conversation will sound choppy, and the quicker the other person will hear you. The following diagram shows you the minimum hardware requirements for an Internet phone setup:

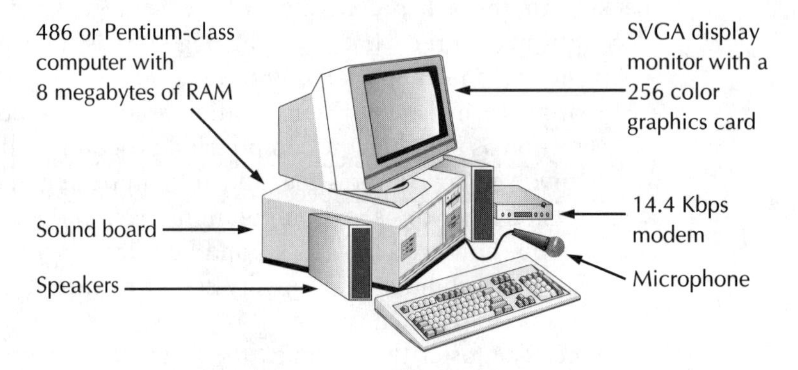

note *Programs like WebPhone, which strive to offer high-quality sound, using complex compression schemes, can* **run** *on 486 computers, but they sound better on faster Pentium-level machines. Pentiums can process information faster, and their faster modems can send more information more quickly. All this compressing and decompressing happens very fast, on-the-fly. As soon as someone starts speaking, the digitization and transmission begins, so the computer had better be quick and the connection had better be fast, or else you might lose a word or two. And different programs achieve different levels of quality by implementing different ways to compress sound.*

Codecs Streamline Your Data for Traveling Light

There are about ten different compression schemes, commonly referred to as *codecs* (short for *compression/decompression*), incorporated into today's plethora of Internet phone software. Some codecs, basically mathematical equations, work better with low-speed connections, while others focus on filling in the dropouts or sound gaps by using a feature called *interpolation*. Still others are really optimized for higher-speed connections, such as 56K and ISDN. Many Internet telephone software manufacturers, such as Telescape, have taken standard codecs and further enhanced them by making them transfer data even faster.

remember *If you're interested in super-fast connections, such as 56K, ISDN, and cable modems, review Chapter 4 for detailed information about what high-speed lines are and what they can offer you.*

Popular Codecs

The most popular codec compression scheme is GSM, short for *Global Standard for Mobile Communications.* Created in Europe for European cellular telephones, GSM is built into WebPhone, NetPhone, and Speak Freely, to name a few. Basically, codecs like GSM take the amount of data needed to transfer normal speech over a telephone line and cut it in half. These complex mathematical equations shrink 8000 bytes of data down to 4000 or 2000, or 1650 or 346 bytes of data. (Now don't you wish you had paid attention in algebra?)

When communicating with someone over an Internet phone, you both have to be using the same codec so that those little sound packets can be read on either end. This is one reason why your buddy who is hooked on WebPhone won't be able to call you if you're an IRIS phone fanatic. If the program you are using doesn't have the ability to understand the codecs used in the program your friend is using, you just can't talk to each other. It's like two people speaking in two different languages. More and more Internet telephone programs allow users to either manually or automatically select different types of codecs. This provides great flexibility when you have a noisy Internet connection.

note *Codecs you might run into include ADPCM, GSM, PCM, and TrueSpeech, to name a few.*

Again, you don't really have to know this stuff to make a phone program work. I can tell you that dear old Dad has no clue what codecs are. But he does know that when the Internet is busy, or when his software tells him he's connected to his service provider at 26,400 baud instead of 28,800, he can expect a crummy connection. He definitely noticed a difference in sound quality when he upgraded from a 486 computer to a Pentium. And when he tests new Internet phone programs he notices the sound qualities vary drastically. Just like with art—Dad may not know codecs, but he knows what he likes.

It really doesn't hurt to delve a little deeper into the various types of codecs. Especially since so many programs, such as PGPfone, Speak Freely, and ePhone let you pick and choose what compression scheme you want to use. And knowing what each does and how it works with the various types of computers and speeds of Internet connections will help you get better Internet phone conversations.

Besides, I can't tell you what a real party-starter talking about codecs can be. Try this next time you find the conversation lagging at your next cocktail party. "Did you know that GSM reduces the data rate factor by almost five?" You'll have 'em eating out of the palm of your hand.

How Codecs Work

Table 5-1 provides information about the most popular codecs, and explains what you can expect in terms of sound quality and how powerful a mathematical workhorse computer you'll need for each. As a point of comparison, keep in mind that uncompressed sound travels over a traditional telephone connection at a rate of 8000 bytes per second, and the sound quality is generally excellent.

Compression	Bytes per Second	CPU Needed	Sound Quality
Simple	4000	486/33MHz	Poor
ADPCM	4000	486/33MHz	Good
Simple + ADPCM	2000	486/33MHz	Lousy
GSM	1650	486/66MHz (but a Pentium 75MHz or above works better)	Good
Simple + GSM	825	Pentium 75MHz	Not so good
ADPCM + GSM	800	486/33MHz	Good
LPC	650	Pentium 90MHz	Good, but depends on quality of connection
LPC-10	346	Pentium 133MHz	Not bad

TABLE 5-1 *Popular Codecs* ■

Nitty-Gritty Codec Information

To satisfy that techno-weenie urge you might have, read on for more detailed background on some of the compression schemes you'll find in Table 5-1:

- **ADPCM** reduces the data rate by a factor of two, essentially reducing 8000 bytes of data per second to 4000 bytes of data per second. ADPCM requires far less computing power than GSM when it comes to encoding and decoding sound. This means that low-range 486 computers, and possibly some 386DX machines, can use the ADPCM compression scheme, provided that the Internet connection is relatively fast.
- **GSM** reduces the data rate by a factor of almost five; in essence, it compresses 8000 bytes of data per second to 1650 bytes per second. However, the encoding is a very complicated process requiring a fairly fast computer in the range of a 486 or Pentium.
- **ADPCM + GSM** cuts the data transfer rate to about 800 bytes per second, but still requires 486-type computer.
- **LPC** reduces the data transfer rate by a factor of 12 using a complex mathematical equation, which means you'll need a floating-point math coprocessor. LPC can be highly sensitive to, clipping which results from too high an audio input level—often caused by setting the gain on the microphone too high or speaking too closely to the microphone. LPC is also quite sensitive to high-frequency sound, so those of you with high-pitched voices may find that it doesn't work for you.

Codecs and Your Computer

You'll notice that some codecs are great for slow computers but require fast Internet connections, while others work well with low-speed Internet connections but require fast computers. Both the speed of your computer and the speed of your modem dictate what type of compression scheme or codec will work best for you. With some programs, like IRIS phone, you can pick a codec to use by adjusting the setting for the speed of your modem connection. Other programs, such as PGPfone, let you actually select the codec by name.

Different codec algorithms compress sound at varying qualities and speed. The better the sound quality, the more computing power it takes to make the transmission flow smoothly. If you use a codec meant for high quality on a slow computer, the sound will break up as the machine tries to keep up with the volume of data it is trying to process.

Why Bother Noticing Which Compression Scheme You're Using?

There is no test at the end of this chapter. You don't need to memorize the speed of each codec. But knowing that codecs should match your machine and modem capabilities helps you understand one of the reasons why your machine may sound better using FreeTel than it would with WebPhone, especially if FreeTel is using a less machine-intensive codec.

tip *If you want to get a feel for how your computer and Internet connection work with the wide range of codecs, I suggest you try SpeakF reely. Speak Freely will let you change the compression settings (codecs) on the fly, and instead of having to have someone on the other end tell you how it sounds, you can use Speak Freely's echo back server to echo back your voice so you can hear it for yourself.*

OK, armed with your newly acquired codec data, you're now ready to hit the party circuit.

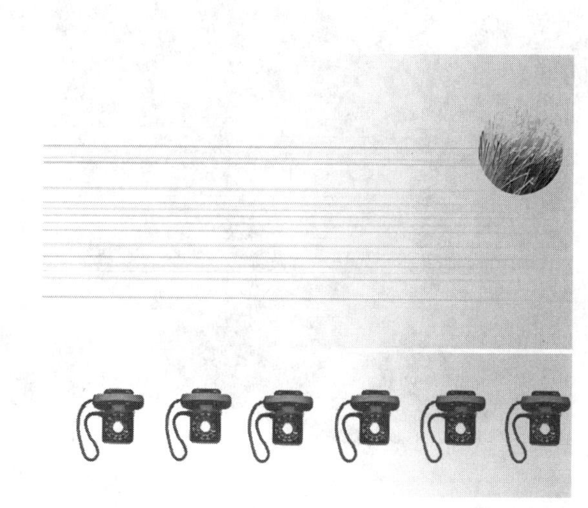

Chapter 6

Internet Phone Software Products

HERE it is. The BIG list. All the Internet phone programs I could find. In this chapter I'll try to give you as much information about each Internet phone program as I can. I've included information about what platforms the programs run on, whether they worked the first time I tried them, and any interesting features I thought were useful to point out.

Amazingly, it seems that every day a new Internet phone program enters the market. Keeping up with it all can be a dizzying experience. So if you want to stay abreast of the latest and greatest Internet phone developments, updates, and new product announcements you can always subscribe to my Internet phone mailing list. It's totally free and is a mix between questions and answers and new product announcements and reviews. Just drop me an e-mail at **netphones@aol.com** and I'll put you on the list. The list is also a great place to find tips and tricks on how to tweak your Internet phone.

How I Rated the Programs

Some phone programs are good. Some are bad. Some are just downright awful. To make it easier for you to quickly spot the best, I've included a rating system for what I feel are the three most important features: sound quality, ease of use, and whether the program worked the first time out of the starting blocks. The ratings for sound quality and ease of use start at one and go all the way up to five, with five being totally primo.

I've tried to be fair in my assessment of the programs, having had numerous conversations with as many people as I could find. I've also tested these programs on both the minimum configuration possible (for a PC that would be a 486/66MHz and for a Mac that would be a Quadra), and the preferred type of configuration (varies with each program), but for me that would be a 100MHz Pentium and a PowerMac. When I rated the voice quality, I took into consideration the network traffic and the relative slowness of my Internet connection here in Alaska.

What you'll find is that most programs offer relatively good sound quality. But the various nuances and glitches each has may mean the program may not work for your type of system or the way in which you connect to the Internet. Some programs rely on directory services, which oftentimes can be so overloaded that it takes literally minutes to connect and find the person you want to call. Others work only with an IP address. So remember, each program will have it's own trade-offs.

Some programs I tested caused numerous errors or just plain didn't work. It was only after a little tweaking that I finally coaxed a couple of these bad boys into performing, and even then some only marginally. I've included my own personal opinion as to why I like or don't like the product in the aptly named "Personal Opinion" section. Again, these are only my opinions, based on my own experiences, my Internet connection, and my computer. I may love Intel Internet Phone, but you may hate it. I may find TeleVox's sound muffled, but it might sound pretty darn good to you, so I highly recommend you try as many as you can that look interesting to you. Then you'll be better prepared to match the perfect program to your system.

Start by picking a few packages that look like they will work with your machine and your Internet provider capabilities. Test them out, then check the manufacturer's home page for additional technical notes and upgrades. Sometimes a small blurb in a technical note will make the world of difference in the operation of a program. Also check the tips section in Chapter 9 of this book for other ways to tweak and twist some of these programs into sounding great.

I could say, "XYZ program is the best of the bunch; don't waste your time with anything else." But that wouldn't be fair to either you or the product manufacturers. You may have an entirely different system than mine, a better connection than I do, and you may like the features I find annoying. So again, this list is a guide to help you figure out which ones *might* work for you and which ones you might want to avoid.

What I Found

First, let me tell you up-front a few of my initial opinions so if you're one of those impatient types you can jump ahead to the Internet phone program that sounds like it might suit you the best.

I personally think Intel Phone has the best sound quality, WebPhone has the niftiest features, and IRIS phone gives you the best voice quality when talking to people overseas. FreeTel is the most reliable for making a decent connection and being able to connect all the time. PGPfone has the best voice encryption. And Speak Freely has an echo-back server that replays any sound you might send to it, so it's an excellent choice if you want to test your microphone, compression, and speaker, plus be able to hear exactly how you sound.

I had the hardest time connecting with DigiPhone due to my own and my Dad's Internet service provider. It simply wouldn't work between our two connections. But in DigiPhone's defense, I did have a couple of clear conversations with people whose providers did work well with DigiPhone. The most system errors came from Web.Max.Phone. Great interface, but maddening system lockups and crashes made it the worst of the bunch as far as reliability.

I wasn't able to connect easily on the first, second, third, or fourth try with Internet Call, although when I did connect, the voice quality was relatively good, especially to the few people overseas I talked with. This is clearly a program in the works, like so many other shareware and freeware Internet phone products.

I found TeleVox to have decent sound quality, but the other person always sounded like he or she was in a tunnel. Also, sometimes the sound would be quiet, but then all of a sudden the voice would blast out when the Internet was overloaded. So be careful when using headphones with TeleVox. It works best when the Net isn't very busy.

Speak Freely and PGPfone have compression settings that aren't what you'd call self-adjusting, so I really had to tweak them to get just the right mix for my computer and slow Internet connection. But when I did, the sound was pretty darn good. In addition, Speak Freely has a simple interface that can be mixed with Phil Zimmerman's Pretty Good Encryption program, and PGPfone offers the built-in ability to encrypt conversations—features which I'd say make these two products the best at offering truly private conversations.

The busiest Internet phone server-based programs, or those for which you're sure to find someone to talk to but may have to wait just to connect to the servers, were VocalTec's Iphone, CoolTalk, Intel's Internet Phone, and FreeTel. Especially in the case of CoolTalk and VocalTec, the overload seemed to affect connect time and sound quality.

The least crowded servers—or should I say those serving the least number of people regardless of the time of day—were SoftFone, IBM's Internet Connection Phone, DigiPhone, and IRIS Phone. These are extremely capable programs, but if you get one you may have a hard time finding people to talk to. Most of them aren't frequently used, due in part to very quiet marketing campaigns on the part of the companies that produce them.

Although these products are listed in Chapter 10, I must say that the most bizarre conversations I had were when I tried video/audio packages such as FreeVue, VDOPhone, and CU-SeeMe. Actually, VDOPhone generated great color pictures and amazingly good sound, FreeVue had the best refresh rate, and CU-SeeMe offered a wide range of varying connections. It wasn't that the programs were at all unusual. The people on the other end, however, managed to find utterly outlandish things to do with a video camera and a connection to the Internet.

remember *If there are children present, do not use these programs to randomly connect to other users and "see what they are doing." Otherwise, the kids may get a lesson in sex education that you don't want them to have.*

All in all what excellence in a phone product really boils down to is whether the program can run on your computer, and whether you want to use it to meet new people or simply talk to Mom and Dad. To meet new people, look for programs that offer server-based connections; for one-to-one conversations, consider those that provide direct connections.

How You Should Approach Testing Phone Software

My best advice is to be patient when you start on your Internet phone testing voyage. Not all will work the first time. Many may bomb out because you don't have your machine set just so. Others just won't work with your Internet service provider (ISP). While still others may require you do a little tweaking by choosing different sound compression schemes. Keep in mind that all this testing *will* take time. Your first crystal-clear connection may happen the first time you call someone or after several days of futzing with the software.

If you aren't adept at computers in general here's some encouraging news. If my Dad can make an Internet phone product work, so can you. Dad started out not knowing how to download files or even how to find them on his computer once he had downloaded them. He was totally clueless about how to configure and use various types of Internet phone programs. With a little prodding, Dad's now the one jumping up and down asking to test the latest and greatest program. He has really come a long way. And if you're beginning to think these products might be too technical for you, think again, and just give one or two of them a try. Once you start using an Internet phone, I think you'll be amazed at how simple it is.

Practice Makes Perfect?

It took us about a day to test just five of these Internet phone programs. That's a long time when you consider testing a real phone takes about a minute. Some programs didn't connect on the first try; others did and sounded great, and some took hours to figure out the best way to configure them for our particular setup.

Practice is what really helped Dad understand what he was doing. And exposure to lots of Internet phone programs helped him understand how the different types of Internet phone programs work but, more important, that not all of them work on the first try, or even work at all.

I think that this was a key turning point for him. At first he *thought* the products should all work with no problem, and when they didn't he thought *he* was doing something wrong. After testing enough he realized that half the time they may simply not work as advertised, and that he shouldn't blame himself for a lousy or nonexistent connection. There are just too many variables, himself excluded.

I'd recommend that if you want to increase your computer comfort level, do what Dad did. Practice. As with anything you want to at least understand and at most master, you have to practice. You have to take time to sit down and concentrate. And more important, you have to know when to give up and say this just isn't working and move on.

remember *A great deal depends on the person with whom you are connecting. If he or she has a slow machine, a bad connection, a cheap modem, or a bum microphone, and is yelling into the microphone, then don't expect to hear every single word said. I recommend you don't base your final decision on which product to use regularly from just one call. Try calling several people at different times of the day. You'll definitely notice a difference in quality especially during different times of the day and between callers.*

If you can't find anyone to talk to, and want to test one of the products, send me an e-mail message at **netphones@aol.com,** and/or drop by my Web page at **http://www.netphones.com**. Feel free to click on one of the phone links so we can talk.

Remember—They're Free!

Above all, remember that you're getting two things: virtually free long-distance calls and the chance to meet people around the world. So don't whine too much if everything isn't perfect. None of these software programs, and almost no Internet service provider, guarantees a certain quality of service like the phone company does. I guess in some part, the old saying, "You get what you pay for," sort of applies to Internet telephony. If you want to save some money, you're going to have to work at it. An Internet phone is nowhere near as simplistic as an ordinary telephone.

Internet Phone Listings

It's assumed you will have the minimum requirements, including at least a half-duplex sound card, a microphone, and a true connection to the Internet. So I won't list whether the program is full- or half-duplex (almost all are both), or if it requires a sound card, microphone, and speakers (they all do). Instead, I'll outline the other computer and operating system requirements each has.

If you are using America Online, Netcom, or CompuServe, be sure to check Chapter 9 for tips and information on what is needed to connect Internet phones with these providers. You *must* have an IP address for these programs to work. And not all connections with major commercial services offer you that. It depends on the version of the software connection software you are using. If you are in doubt as to whether these programs will work with your specific provider, just ask the provider. He or she should know.

Explanation of My Ratings

VOICE QUALITY How close to a regular telephone did it sound?

EASE OF USE Was it straightforward and easy enough for a very casual Macintosh or Windows neophyte to use?

OVERALL PRODUCT FEATURES When you add up all the features and add the sound quality how does it rate?

WORKED THE FIRST TIME? Was I able to make a call and talk to another person the first time I ran the program?

SERVER-BASED? Does the program need to connect to a server in order to operate or can you call someone directly either through their e-mail or IP address?

TEXT CHAT? Does it offer the ability to type messages to the other person (a useful feature when the connection may be choppy)?

FILE TRANSFERS? Can you send a computer file directly to the other user while you two talk?

VOICEMAIL? Does it offer the ability to record a message and send it to the other party when they don't answer your call? Voicemail is classified two different ways—online and off-line. Online means the other party is running the program but not answering your call. Off-line means the other party isn't running the program, so your message is sent to a central server or to their e-mail account, and that same voicemail message can later be retrieved by starting the same Internet phone program used to send the voicemail.

COST Is it free? If not, what does it cost? Costs vary, so I've provided a range, such as under $50. Pretty much all Internet phone programs offer time-limited demonstrations, full working programs, but limited in terms of either the amount of time you can talk to another person or in the number of days you can use the program. Others may give away the program in hopes that you'll decide to step up to the pay version.

LINKABLE TO A WEB PAGE? Can you dial someone directly by clicking a link from their Web page?

NIFTY FEATURES What feature or features separated this program from all the rest or provided something unique to the user?

The Big List of Products

Intel Internet Phone

Intel Corporation
2200 Mission College Blvd.
Santa Clara, CA 95052-8119
Home Page: http://www.intel.com
E-mail: info@intel.com
Platform: Windows 95

Ratings

Voice Quality: ☎☎☎☎☎
Ease of Use: ☎☎☎☎
Overall Product Features: ☎☎☎☎✆
Worked the First Time? Yes, but the directory listings can be confusing.
Server-Based? Yes, but you can also call if you know the person's IP address.

Text Chat? No
File Transfers? No
Voicemail? No, but does offer option to see who called you while you were running Intel Phone but not answering calls.
Cost: Free; still in beta version 2.0 at time of writing.
Linkable via Web Page? Yes
Nifty Features: Includes the Intel Connection Advisor utility, a little program that runs while Intel Internet Phone is running, showing you if traffic is heavy or if your machine's CPU capabilities are being taxed.

Minimum Requirements

- 90MHz Pentium
- 5MB free disk space
- 16MB RAM
- Microsoft Internet Explorer 2.0 or Netscape Navigator 2.0 if you want to use the directory listings to find other Intel Internet Phone users. You can also use it stand-alone to direct-dial someone's IP address.

note *Intel Internet Phone can work with a 486/66MHz machine, but it is relatively slow. Some non-Intel computer users have reported problems. The best thing to do is try it if you have a 486 or a non-Intel-compatible system.*

Also, Intel reports that AOL users cannot use Intel Internet Phone, because AOL does not provide a true IP address, and that it cannot work with InternetMCI's TCP/IP stack.

Features

Intel Internet Phone is one of the first Internet phones to conform to the industry standard H.323 compression, meaning other Internet phone programs that use the same standard can call Intel Internet Phone users, regardless of their operating system.

The program can run with or without a Web browser (see Figure 6-1). However, it does use the Web browser (either Netscape or Microsoft Internet Explorer) to help you find other Intel Phone users who have downloaded and are currently using the program. It will automatically launch your browser if it's not already running when you launch Intel Phone, and take you to the directory service you have specified in the settings of the program.

You can also use the Web browser to display the names of people who have called while you were logged into the Net, running Intel Phone but not available to answer calls, and to quickly jump to those people you call most frequently through the use of the HTML-based Quick-Dial phone book.

You can also quickly search for and e-mail other Intel Internet Phone users using the Web-based directory services. Plus, you can run it while you are running EasyPhoto, an application that lets you exchange photographs during your Internet phone call. The only negatives are that it has no text chatting feature, lacks voicemail, and provides no shareable whiteboard. It's strictly for talking to other people, period.

Half-duplex users are offered a "push-to-talk" button; full-duplex users get automatic voice activation. Both can adjust the incoming volume or mute their conversation.

Personal Opinion

Intel Phone by far has the best overall sound quality, provided you have a fairly fast machine and a relatively good Internet connection. I've only had one really poor conversation and that was during a period of countrywide network problems.

Speech delay is almost nonexistent, which makes conversations on Intel Internet Phone sound more like real telephone conversations than most of

FIGURE 6-1 Intel Phone and Intel's home page

the other programs. There's no need to adjust compression rates or go through elaborate configurations. That's all handled by the software.

Its other outstanding feature is the interface. It's extremely easy to use once you've figured out how the directory service portion works. That was the only truly confusing part, and something that confused not only me, but almost everyone I talked to when the product first became available. Now that the directory services have made it a little easier, by providing links to users who are currently online and using the program, some of that confusion has gone away. If you plan to use the Direct Dial feature, like I do to talk to Mom and Dad, you'll need to know their IP address. For me it's a little cumbersome to have to get my IP address, then e-mail it to the folks to let them know where to dial—so much so that I've finally broken down and paid for a more expensive static IP address. Sure we could both just go to the directory services, but it's much simpler and faster for me to pay the extra bucks, and better for Mom because she knows she can dial me up anytime and I'll be there to answer the phone.

The best thing about Intel Phone is that the interface doesn't get in the way. It's simple, elegant, and not so feature-packed you are overwhelmed. It's great for someone like Mom, who just wants to talk. No extraneous bells or whistle to confuse novice PC users. Obviously, a lot of thought was put into making the product unobtrusive and integrated into your Internet screen. It's just a single bar, which can attach itself to your Web browser or float independently anywhere on your screen.

If you have a Pentium, try this one first. I guarantee you'll be impressed.

FreeTel

FreeTel Communications, Inc.
540 N. Santa Cruz Ave., Suite 290
Los Gatos, CA 95030
Fax: (408) 358-6385
E-mail: freetel@freetelco.com
Home Page: http://www.freetel.com
Platform: Windows 3.1, Windows 95, Windows for Workgroups

Ratings

Voice Quality: ☎ ☎ ☎ ☎
Ease of Use: ☎ ☎ ☎ ☎ ✆
Overall Product Features: ☎ ☎ ☎ ☎
Worked the First Time? Yes, with absolutely no problems.

Server-Based? Yes. No direct IP connections.
Text Chat? Yes
File Transfer? Yes
Voicemail? No
Cost: Free—supported by advertising.
Linkable via Web Page? Yes
Nifty Features: Identifies callers engaged in conversations with an asterisk (*), making it easier to spot people who are available to talk.

Minimum Requirements

- 486/33MHz PC
- 4MB RAM
- 1MB free disk space

note *FreeTel is one of the few Internet phone programs that will work with a Unix host account using TIA (The Internet Adapter) or SLIRP to simulate PPP or SLIP. If you have a shell account on a university computer you might be able to use FreeTel. Check with your system's administrator for instructions on how to set up the TIA with your university computer.*

Also, some ISPs block transmission of FreeTel packets. If you can't connect to the server, check Chapter 9 for the information you'll need to give your service provider.

Features

Like Intel Phone, FreeTel is very simple (see Figure 6-2). Since it's server-based and automatically connects to the server once you launch the program, you don't have to know your IP address or your e-mail server host name. The program handles all that.

The directory is fully searchable by name and will display a list of users who are currently using FreeTel based upon information you supply in your search criteria. The directory also alerts you to those who are already in conversations by placing asterisks before their names. That way you know exactly who is available to take your call and who isn't.

The program does all of the compression negotiating to provide you with the best possible audio. This is done by adjusting the compression schemes based upon the speed of your modem. Therefore, it's critical you supply the correct information in the initial configuration screen when you first run the program. If the Net is running slowly and chopping up your conversation

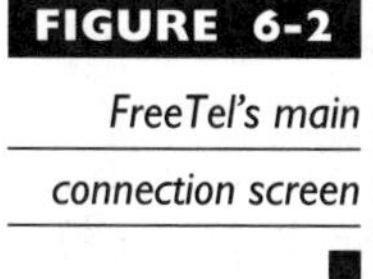

FIGURE 6-2

FreeTel's main connection screen

you can always invoke the Booster feature, a software switch that tries to cut out the choppiness, but at the same time can slow down your conversations by increasing the delay.

Families with kids or businesses with several users can create multiple configurations. Each of these configurations can have its own text-based greeting message and unique login name. You don't have to go by your e-mail address, your first name, or your initials. You can use any alias you wish.

Other features include the ability to transfer files in the background while you talk, along with the ability to place a link on your Web page so people can call you, provided you are running FreeTel at the time.

Text chatting is also available. The text you type will be displayed immediately with no need to press ENTER. Text chatting is a great feature when you need to relay a URL address or when the voice quality is marginal, or nonexistent. With it you can still communicate in real time.

The program is sponsored by advertising, which appears as a banner at the top of the program's main window. These ads are far from obtrusive, but if you want to rid yourself of such commercialism, for $29.95 you can purchase an upgrade to the product that allows you to remove the ads.

Personal Opinion

Simple, not fancy, but rock-solid and reliable. That's the best way to describe FreeTel. Like Intel Phone, it's a single-screen product with all the features, including text chatting, volume controls, and file transfers—all accessible at the touch of a button via the main window.

I've never had the software bomb out on me, nor had a single problem configuring it. You don't even need an e-mail account, because FreeTel configures itself to its own servers. It's a true no-brainer installation.

The voice quality isn't as good as Intel Phone, but its close. I've only had one really bad conversation on FreeTel, with a fellow in Australia, but again, the Net was very busy at the time. And even then the text chat feature came in handy and we ended up text chatting for about an hour once we realized neither of us could make out what the other was saying.

When Dad still had his lowly 486/33MHz portable, FreeTel worked like a champ, especially compared to other programs that required heftier machines. Although the sound delay was sometimes frustrating we could always understand each other and had few dropouts in our conversation. The text chat feature made it easy to enhance our conversation when we had critical information to exchange, such as the latest URL for the Oscar Meyer Wienermobile.

If you haven't made the plunge yet and don't have the muscle to run Intel Internet Phone, definitely check out FreeTel. I'd even recommend downloading it and keeping it on your machine as a backup in case the other programs don't work or their directory services are busy. I do.

The only problem with FreeTel is that it lacks sophisticated features such as voicemail and application sharing. And, compared to the Intel Internet Phone's sound quality, FreeTel's sound is more like AM radio quality, whereas Intel Phone sounds more like a true telephone. And sometimes the voice delays can be maddening. But the company vows to keep FreeTel free, and at that price you really can't go wrong.

Web.Max.Phone

Berkeley Systems
P.O. Box 9749
Berkeley, CA 94709
Fax: (510) 540-4708
E-mail: info@berksys.com
Home Page: http://www.berksys.com
Platform: Windows 95

Ratings

Voice Quality: ☎ ☎ ☎
Ease of Use: ☎ ☎ ☎ ☎ ☎ (If it works)
Overall Product Features: ☎
Worked the First Time? No. Neither the real version nor the demo worked on my machine or five other of my tester's machines.
Server-Based? No. Direct dial via e-mail or IP address.
Text Chat? No
File Transfer? Yes
Voicemail? No
Cost: Under $40 when purchased with Web.Max system
Linkable via Web Page? No
Nifty Features: None

Minimum Requirements

- 486/66MHz PC
- 8MB RAM
- 5MB free disk space for just the phone program; 18MB for the entire Web.Max system

Features

Web.Max.Phone (see Figure 6-3) is simple to configure, and can work with or without its companion products Web.Max.Retriever, Web.Max.TV, and Web.Max.Security. As a matter of fact, the only information needed to configure Web.Max.Phone is your e-mail address and your name. Everything else is taken care of by the negotiating server you connect to when you bring up the program. You can, however, include a picture that will be transmitted to the other user's Web.Max.Phone screen when you establish a connection.

There are literally no compression settings, CPU settings, or voice controls—just a few buttons to add and delete people's names and e-mail addresses to your phone book, and a big Talk button to press when you want to speak. You simply add a person's name and e-mail address to your list, then click on the dial button and the program does the rest. An indicator

FIGURE 6-3 Web.Max.Phone's simple interface

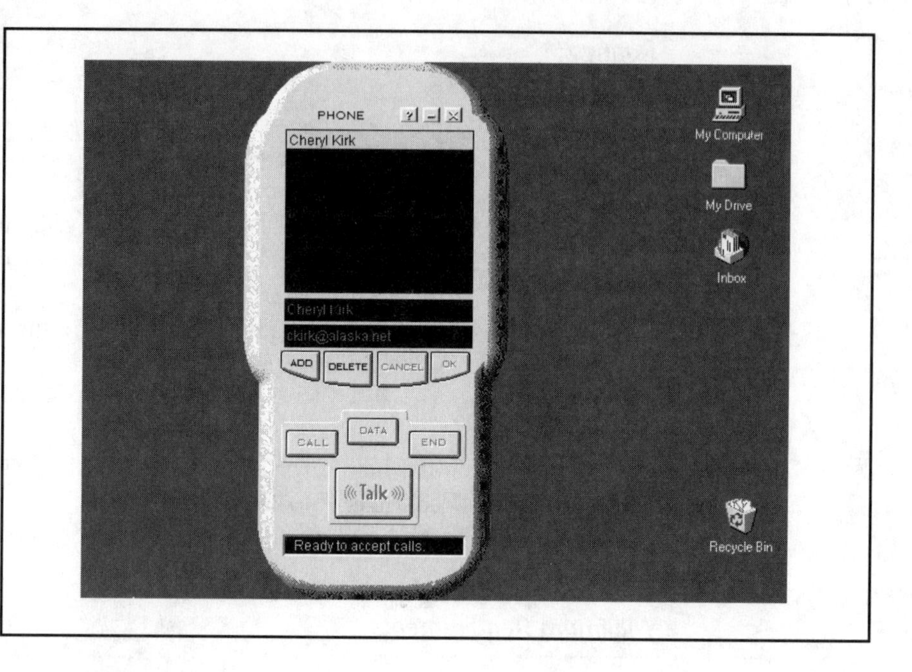

at the bottom of the screen informs you when someone is calling, the status of your call, and the status of any file transfers you are performing.

Unfortunately, there is no directory of online callers so you have to know someone who already has the program. But once you do, using it is as simple as inputting their e-mail address and pressing dial. The phone automatically picks up when someone calls.

Personal Opinion

Have you ever heard the saying, "It's not what you know, it's how you look, that really matters"? That's the best way I can describe this program. It's so easy. Such an elegant interface, but after trying to connect with four different people I finally gave up.

I did have one decent very brief conversation for about two seconds. Then the program froze up and shut down my computer. I finally gave up after the tenth try, after it had managed to completely eat my sound driver.

When it did work that one time, the sound of the caller resembled that of someone talking in a empty trash can, kind of a distant sound somewhat akin to talking to someone on the moon. Although I like the interface, I just wasn't impressed with the usability of the program. It's the least desirable of the lot and unless you just want to experiment with it, I'd stay away from it.

TS Intercom

Telescape Communications Corporation
1965 West 4th Avenue, Suite 101
Vancouver, B.C. , Canada 1M8
Fax: (604) 469-5589
E-mail: info@telescape.com
Home Page: http://www.telescape.com
Platform: Windows 3.1, Windows for Workgroups 3.11, Windows 95

Ratings

Voice Quality: ☎ ☎ ☎ ☎
Ease of Use: ☎ ☎ ☎ ☎ ✆
Overall Product Features: ☎ ☎ ☎ ☎
Worked the First Time? Yes
Server-Based? No. You can connect by simply typing in the person's e-mail address.
Text Chat? No
File Transfer? Yes
Voicemail? No, but you can send e-mail from within the application.
Cost: Free
Linkable via Web Page? Yes
Nifty Features: Ability to launch Web browser and e-mail applications directly from TS Intercom Lite

Minimum Requirements

- 486/33MHz PC
- 4MB RAM
- 2MB free disk space

Features

TS Intercom (see Figure 6-4) offers point-to-point communications with a software product that you can distribute freely to your friends and relatives. You can call another person by simply typing in his or her e-mail address. You don't have to worry about connecting to servers or knowing IP addresses. Or you can store their names in your phone book and use that to connect to frequent callers.

FIGURE 6-4

TS Intercom's main screen

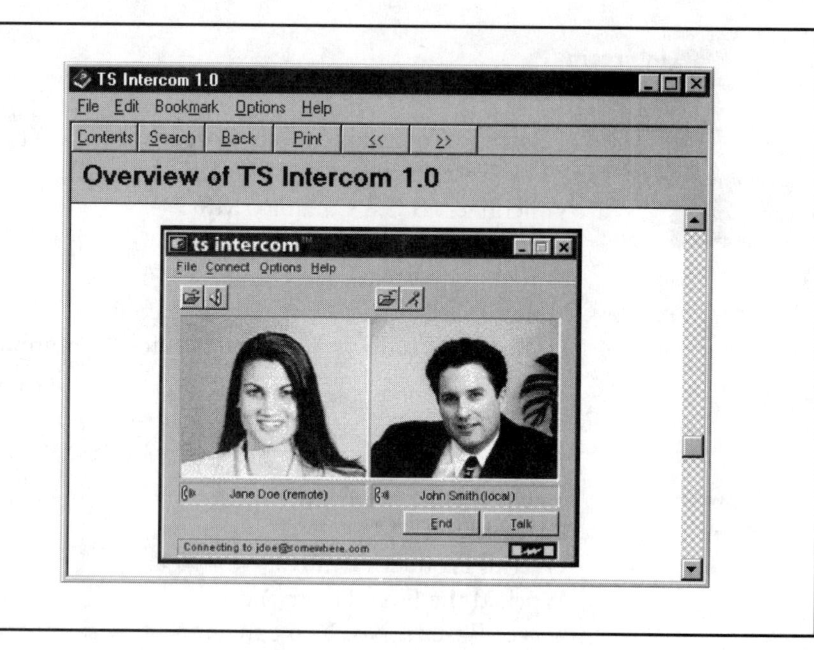

With TS Intercom's Viewport option, you can display a photograph or any graphic file to the person you're talking to, while you're talking. You can also send and receive computer files and e-mail right from TS Intercom while talking to the other person. The e-mail feature links directly to your e-mail program, bringing it up when invoked. Both the sound and microphone levels are fully adjustable during conversations to ensure the clearest possible sound.

note *There is a professional version of this product available. The product reviewed and included on the CD is TS Intercom Lite.*

Personal Opinion

There isn't much to this program. You just bring it up, type in an e-mail or IP address, and away you go. You just can't beat it for simplicity.

The sound quality is relatively good most of the time, although a little muffled at other times. The nifty feature of being able to see a picture of the caller is nice, so having a large enough screen to see it helps. And being able to transfer files in the background while talking makes this an excellent program for small companies that need to send computer files back and forth easily.

It's so bare-bones that you almost think to yourself, Is this all there is? But when you consider that the main function of any Internet phone program is to transmit your voice in real time, and do it clearly, then TS Intercom does that and does it well.

The company's approach seems to be that if you want other features use the other Internet programs you are already familiar with. And TS Intercom will help you do that by allowing you to quickly launch your Web browser or e-mail client directly from TS Intercom, something most Internet phone programs don't let you do. So if you are happy with your browser, content with your e-mail program, and just want to add voice communications, take a good look at TS Intercom.

IBM Internet Connection Phone

IBM Corporation
E-mail: askibm@info.ibm.com
Phone: (800) 426-4968
Home Page: http://www.Internet.ibm.com/icphone
Platform: Windows 95

Ratings

Voice Quality: ☎ ☎ ☎ ☎✆
Ease of Use: ☎ ☎ ☎ ☎ ✆
Overall Product Features: ☎ ☎ ☎ ☎
Worked the First Time? Yes
Server-Based? Yes
Text Chat? No
File Transfer? No
Voicemail? No
Cost: Will be provided with IBM products in the future. Currently free in beta.
Linkable via Web Page? Yes
Nifty Features: Adjustable, "sliding" voice quality control

note *If you have an IBM computer with an Mwave connection, the MWave version of this product will work with no problems—one of the few Internet phone programs that will. Make sure you install the correct version for the setup you have.*

Minimum Requirements

- 486/66MHz PC (although it worked fine on a 33MHz machine)
- Comes in versions compatible with plain vanilla computers or those utilizing the Mwave modem connections.
- 8MB RAM
- 5MB free disk space

Features

IBM Internet Connection Phone (see Figure 6-5) is entirely server-based, showing only those users currently running the program. At the time of this writing, it worked off two different servers, which were never very active.

Once you select the server and the person you want to call, you can adjust the sound compression on the fly as you speak, and you can make changes to the microphone and volume controls with a simple sliding lever.

You can view and call back any previous callers who tried to connect when you were talking to other IBM Internet Connection Phone users. Plus, you can see the name and/or the e-mail of connected users by clicking on their names listed in the server.

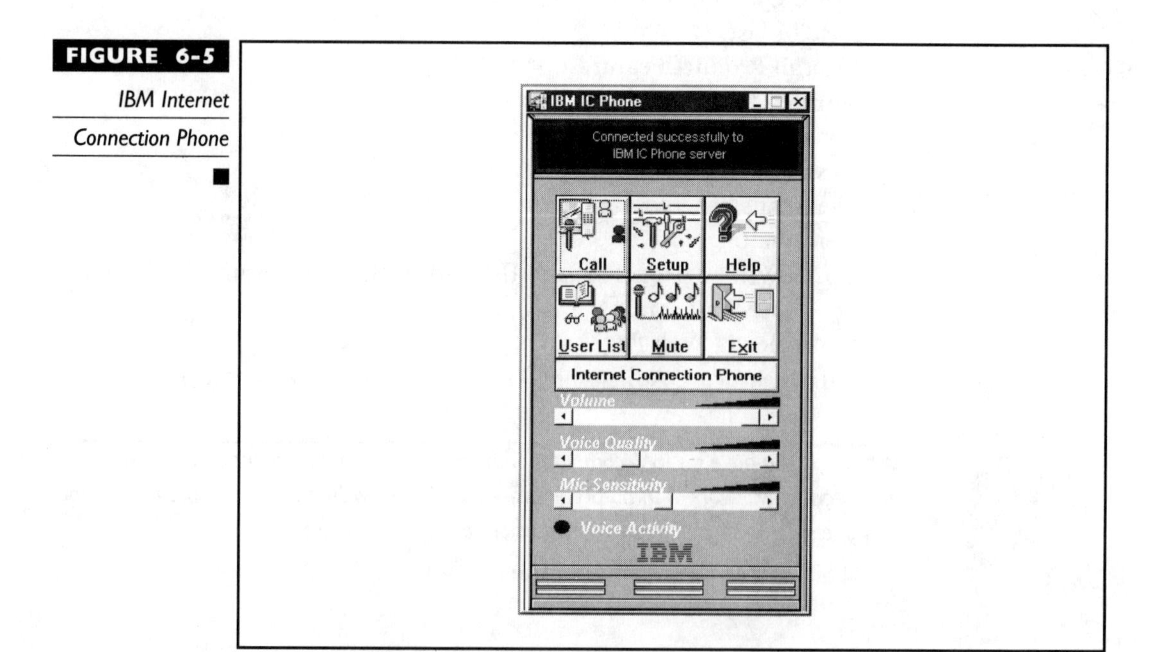

FIGURE 6-5 IBM Internet Connection Phone

The interface is very simple, offering six big buttons to accomplish pretty much all the tasks needed to make the program work. You only have to supply your name and e-mail address. IBM Connection Phone takes care of the rest.

IBM has big plans for the product and is going to include it on pretty much all of the hard drives of new computers they sell. They also plan to link it to telephone systems, allowing non-IBM Internet Connection Phone users to be able to log on and call IBM Internet Connection Phone users.

Personal Opinion

I think this is the sleeper of the Internet phone programs. It's interface, although clunky (IBM still seems to operate in that cheesy low-resolution graphics mode), is simple to understand and easy to manipulate. And the voice quality is very, very good. I'd compare the quality to that of a standard cellular telephone. There is none of that "tin can" sound some Internet phone programs have.

The best thing about it is that no one seems to know about it, meaning the servers aren't clogged and overflowing like VocalTec's Internet Phone. But that's also the worst thing about it when all you want to do is try the program out.

Getting it up and running took a total of seven minutes, including the time it took to download the program, so don't worry about having to futz around with server names and compression rates. The only problem is that it lacks features. Although you can easily adjust the voice quality without having to know anything about compression codecs, it has no text chat, something I think every phone program should have, just in case you get a bad connection and want to relay information to the caller without having to hang up and call again.

But although it lacks any voicemail or file transfer features, the program, like IBM, the slow-moving giant, will most likely evolve over the years into an integrated feature product included on every new IBM computer's hard drive.

Internet Call

Centre for Internet Exchange Technologies
Ho Sin Hang Engineering Building, Rm. 724
Department of Information Engineering
Chinese University of Hong Kong
Shatin, Hong Kong
Phone: (852) 2609 8445; (852) 2609 8385

Fax: (852) 2603 5032
E-mail: cmwu@cixt.cuhk.edu.hk
Home Page: http://dsp.ee.cuhk.edu.hk/proj/icalldl.html
Platform: Windows 95, Windows NT

Ratings

Voice Quality: ☎☎☎
Ease of Use: ☎☎☎
Overall Product Features: ☎☎☎
Worked the First Time? No. It wouldn't find the domain I specified or return any list of users within the domain.
Server-Based? No. Direct dial to the IP address.
Text Chat? No
File Transfer? Yes
Voicemail? No
Cost: Free
Linkable via Web Page? No
Nifty Features: None

Minimum Requirements

- 486DX2/66MHz PC
- 8MB RAM
- 1MB free disk space

Features

There's almost nothing to this program (see Figure 6-6). A single screen displays a pink telephone you click on to initiate a call. At that point you supply the IP address or the actual domain address of the person you are calling.

The program uses a CELP voice coding algorithm to compress a 128 Kbps PCM voice signal to only 4.4 Kbps. This essentially means you can have conversations with people very far away from you on very slow Internet connections and still be able to hear them without much delay.

Interestingly, Internet Call is a free product developed by the department of electronic engineering at the Chinese University of Hong Kong. The university has plans to offer its companion conferencing program, Gather-Talk, to the local phone companies in Hong Kong. So you should definitely

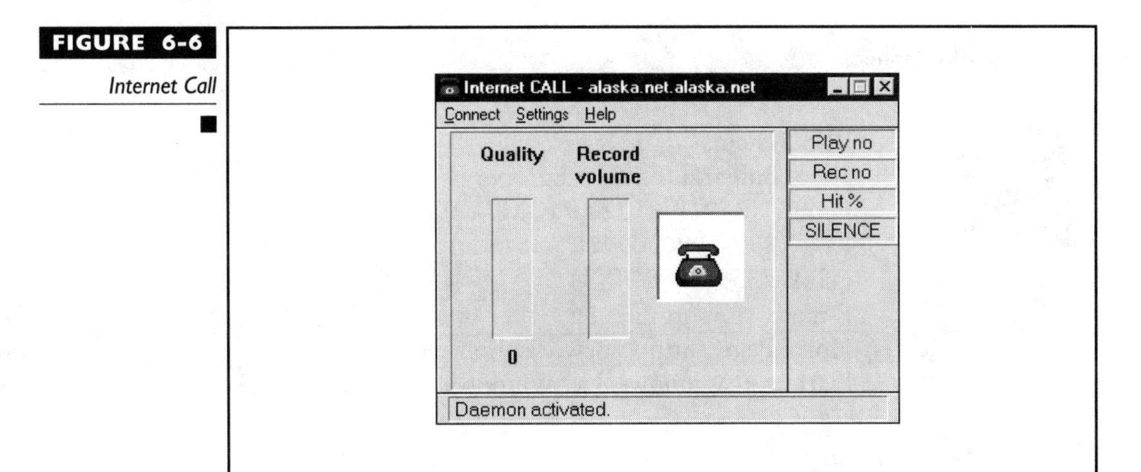

FIGURE 6-6 Internet Call

see GatherTalk develop, and hopefully its little brother, Internet Call, will also get some attention.

Personal Opinion

The sound quality is really not that bad, but getting connected didn't work the first, second, or third time. For some reason, network traffic possibly, it couldn't find the intended user even though I input the correct IP address and domain name.

Once I did find someone to connect to, the sound was much like a CB radio—nothing spectacular but still very audible. I'm sorry I can't jump for joy about this product; nothing about it is that exciting, including the interface, which is very rudimentary.

However, the Hong Kong government seems to strongly back the product and it's cousin, the Internet phone conferencing program GatherTalk, so you should see further development with Internet Call in the future.

Although this program is a no-brainer, I think novices will have a hard time trying to figure out why the program won't connect, a consistent problem for me, and something I solved after having my ISP tweak a few things. Since the program really needs everything to be set just so on your service provider's side in order to work, and since you need to know the IP address of the person you want to call, most people probably shouldn't waste their time trying to get the program to work. Unless you have a really slow Internet connection and are planning to talk to someone very, very far away, the program is probably more frustration than its worth.

IRIS Phone

IRIS Systems
Sofia, Bulgaria (Eastern Europe)
Phone: (359-2) 650 648 (GMT+2)
Fax: (359-2) 9630091
GSM: (359-88) 524884
E-mail: support@iris.bg
Home Page: http://www.irisphone.com
Platform: Windows 3.*x*, Windows 95, Windows NT 3.5*x*

Ratings

Voice Quality: ☎ ☎ ☎ ☎ ✆
Ease of Use: ☎ ☎ ☎ ☎ ✆
Overall Product Features: ☎ ☎ ☎ ☎ ✆
Worked the First Time? Yes
Server-Based? Yes
Text Chat? Yes
File Transfer? Yes
Voicemail? Yes
Cost: Free
Linkable via Web Page? No
Nifty Features: Ability to record your conversations and save them for later playback

Minimum Requirements

- 486/40MHz PC
- 4MB RAM
- 2MB free disk space

Features

IRIS Phone is a feature-packed program, to say the least (see Figure 6-7). It comes with a phone book, voicemail with the option to record different outbound greetings for different people, the ability to record conversations, fairly detailed call tracking, special call-processing features such as different

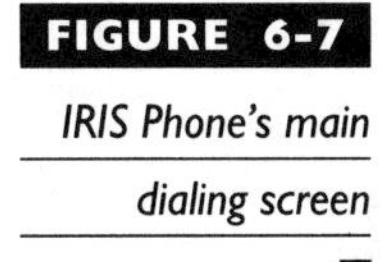

FIGURE 6-7

IRIS Phone's main dialing screen

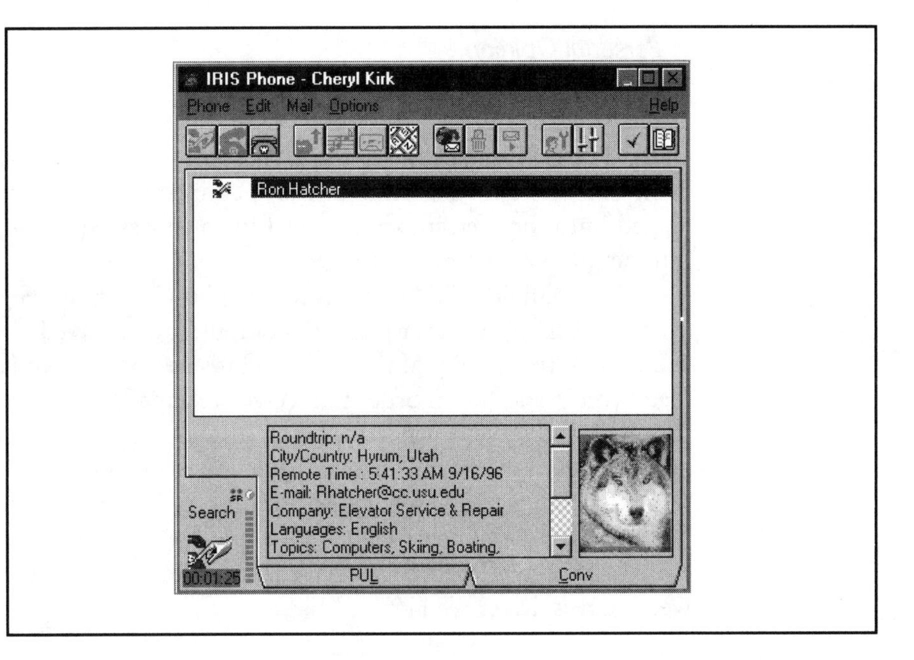

rings for different callers and, during the installation process, the ability to recognize your operating system and install the right version in order to take advantage of your system's capabilities.

The interface is simple and offers a single button to access most of the features. The program can run in the background, allowing you to work on your computer and still take calls and record voicemail messages.

To use the program, you connect to the IRIS Phone server, then select someone from your phone book or find someone to talk to from the list of active users attached to the server. If the user you are calling has included a picture in her or his profile, you'll see that and additional information before you connect.

The program is designed for slow modems and connections to people far away from you. When you get a bad connection, you can adjust the modem speed/compression on the fly, much like IBM Connection Phone, while you talk to other people. The sound quality is very good once you've adjusted the initial settings to match your configuration.

The voicemail feature lets you save messages, record special greetings, and organize voicemail according to caller. Each time you bring up IRIS Phone and log in to the server, your voicemail is automatically fetched for you, and an audio message alerts you to the new voicemail.

You can also transfer files or text chat during your phone conversations.

Personal Opinion

Great. Simply great. The sound quality is very good and the interface very clean. The servers are never busy. And the program installs itself in your Windows toolbar and automatically connects to the server once you've logged onto the Net, making it ideal for business people who want quality Internet phone communications.

And voicemail! What a bonus to any phone program. Since the messages are recorded on the other person's computer, then saved and transmitted as sound files, the quality of the voicemail is superb. Plus, you don't have to be logged on to the Net in order to receive voicemail. All the voicemail files are stored on the server until that person logs in and retrieves them.

You could use it just for voicemail if you really didn't want to bother with trying to connect to someone at certain times. The addition of a text chat board and the ability to send files in the background make it literally the most feature-packed Internet phone program of the bunch. And definitely worth a test drive.

WebPhone

NetSpeak Corporation
902 Clint Moore Rd., Suite 104
Boca Raton, FL 33487
Phone: (561) 997-4001
Fax: (561) 997-2401
E-mail: info@netspeak.com
Home Page: http://www.netspeak.com
Platform: Windows 3.1, WFW 3.11, Windows 95, Windows NT, WIN-OS/2 Warp

Ratings

Voice Quality: ☎☎☎☎☎
Ease of Use: ☎☎☎☎✆
Overall Product Features: ☎☎☎☎☎
Worked the First Time? Yes
Server-Based? Yes, although you can call using an e-mail address too.
Text Chat? Yes
File Transfer? No
Voicemail? Yes

Linkable via Web Page? Yes
Cost: $69
Nifty Features: Multiple telephone lines; hold button; on-hold music

Minimum Requirements

- 486DX/33MHz PC
- 4MB RAM
- 5MB free disk space

Features

WebPhone's interface looks just like a cellular phone (see Figure 6-8), but acts like a full-featured phone system. It offers four lines for incoming calls, call holding with on-hold music, muting, plus Do Not Disturb and call-blocking features.

It also offers a text chat board, the ability to conference multiple callers together, and call transferring. You can dial people directly through their e-mail addresses or use the server to locate someone to talk to in a specific country. If you know their IP address you can use that to call them as well. If you don't know their e-mail address or they have a dynamic IP address

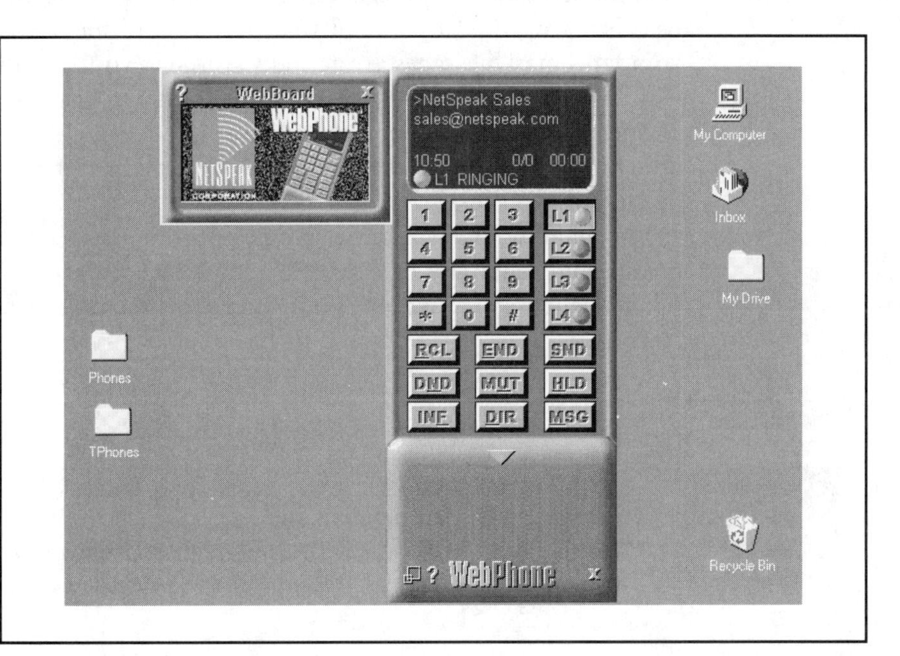

FIGURE 6-8 *WebPhone looks just like a cell phone.*

that changes all the time, you can simply look up their name on the WebPhone server. The first time you connect to a person, you can store their name in your address book, then dial them up from the Web page the next time you call.

The voicemail option lets you set up individual greeting messages for different people, and when you do receive voicemail, the message is sent to your e-mail box, which in turn notifies you of the call.

Personal Opinion

This is by far the sexiest-looking Internet phone program around. It looks just like a cellular phone and acts just like one too. Even though some of the buttons can be confusing for the first-time user, once you've taken the program for a spin, you may not go back to any other.

The voice quality is crystal-clear, because the program does such a good job of taking out the static and background noise. The amount of delay and choppiness is relatively low, but again is dependent upon the speed of your machine and the amount of traffic on the Internet.

If I were a businessperson wanting to sell products on the Web, I would buy WebPhone and link it to my Web page so I could take orders right on the Internet. NetSpeak demonstrates how well this technique works by using it for product support. I've called several times and can vouch for its effectiveness from the customer's standpoint.

Although I like the voicemail features of IRIS a little bit better, you can't beat being notified by e-mail that a voicemail message has arrived. All your voicemail messages are actually stored in your e-mail account, which makes for fast transferring of messages to the program, compared to IRIS phone. The only problem is that if your ISP doesn't allow storage of large attachments in your e-mail account, you might not be able to get your WebPhone voicemail.

But again, with any Internet phone program it all comes down to sound quality and reliability of connecting. On both points this product is top-notch. Definitely try WebPhone if you have a relatively fast computer.

Speak Freely

John Walker (author)
E-mail: kelvin@fourmilab.ch
Home Page: http://www.fourmilab.ch/netfone/windows/
Platform: Windows 3.1, Windows 95, Windows NT version 3.51

Ratings

Voice Quality: ☎ ☎ ☎ ☎ ✆
Ease of Use: ☎ ☎ ☎ ☎
Overall Product Features: ☎ ☎ ☎ ☎ ✆
Worked the First Time? Yes, although you may need to adjust some settings first.
Server-Based? No, but you can log on to a name server to see other users.
Text Chat? Yes
File Transfer? No, but you can send sound files.
Voicemail? Yes
Cost: Free
Linkable via Web Page? No
Direct Dial via IP Address? Yes
Nifty Features: You can listen to how you sound to the other party by using echo servers, which record your transmission and then play it back to you.

Minimum Requirements

- 386/33MHz PC, but a 486 or above is recommended
- 4MB RAM
- 1MB free disk space

note *Your network administrator may have to open up a port to allow Internet Real Time Protocol packets across the network.*

Features

Speak Freely isn't fancy, offering a mainly text-based interface and adjustable sound quality that can rival some of the commercial programs.

You can search an online directory of other Speak Freely users, or you can dial the intended caller directly through their IP address. You can also send a bitmapped picture of your face that will appear on the other caller's screen once you connect, so they can (or at least get a pretty good idea of) who they're talking to.

If the Pretty Good Privacy encryption program is installed on your machine, you can bring it up and exchange your special PGP encryption keys to the other caller when you first connect to him or her. Once you've exchanged your keys, you can set up a secure channel where the two of you can talk without fear of other's eavesdropping on your conversation.

You can also use Speak Freely to talk to UNIX users, and you can broadcast sound to multiple listeners provided you have a fast enough local network. It also allows you to talk to other people who are using Internet phone programs that utilize the Internet Real Time Protocol, or who use the VAT (Visual Audio tool) Protocol, which presumably means those using Maven or ePhone for the Mac.

Speak Freely's most unique feature is its ability to connect to an *echo server* (see Figure 6-9). Echo servers are simply computers that record your voice and then transmit it back to you ten seconds later. The advantage of being able to connect to echo servers is that you can adjust your voice, microphone, and sound settings before connecting to other callers. You can also hear how the different voice compression schemes work with your computer and Internet Connection.

It offers the complete push-to-talk and voice-activated features of most phone programs and works in full- or half-duplex mode.

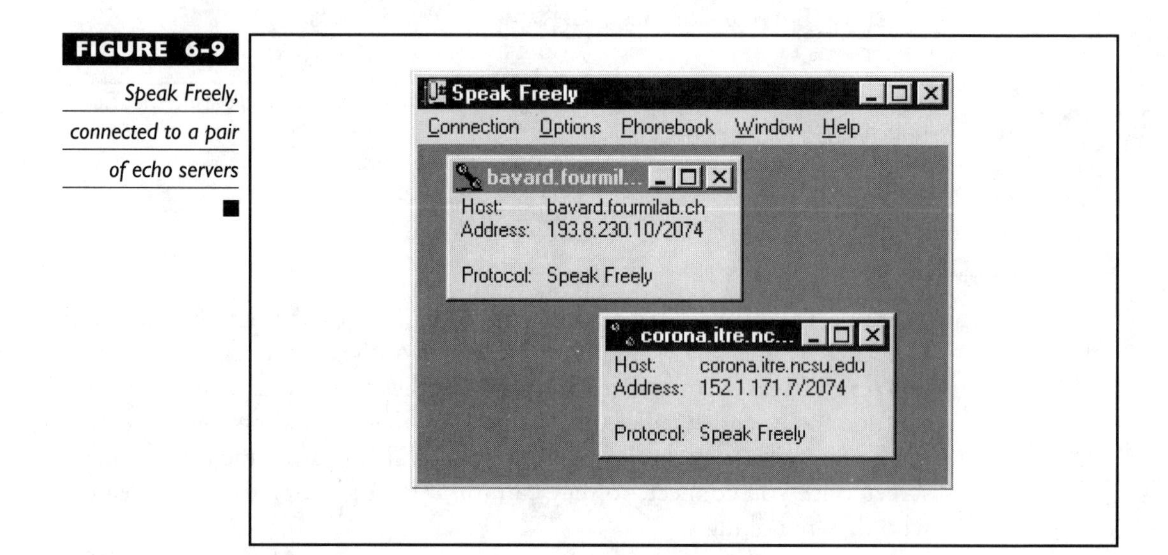

FIGURE 6-9 *Speak Freely, connected to a pair of echo servers*

Personal Opinion

It's not beautiful, but sometimes your favorite tools are more functional than fancy. It's a little sparse in the interface section, but then again, this is the type of program that reinforces the fact that it *is* what you know, and *not* how you look, that really matters.

Although just about any nontechnical person can get it up and running in a matter of minutes, to really appreciate this program is to understand that, unlike programs such as Intel Internet Phone, which don't let you change much of anything, Speak Freely lets you tinker with virtually any setting you please. If you like to tinker you'll like Speak Freely because you really need to tinker with the default settings before the sound is perfect.

I wouldn't recommend everyone use Speak Freely because of this fact. Most people want something to work right out of the box, and unless you *do* tinker with Speak Freely, testing out your system, running benchmarks to find out which compression schemes works best with your connection, changing the protocols you use to transmit voice, listening to your voice echo back from one of the echo back servers, you'll probably be disappointed with the program. And you'd do the program a disservice by not understanding how useful the somewhat technical information can be.

Other programs may sound better on the first pass, but they may not let you make changes on the fly when conditions change, like Speak Freely does. And that flexibility is invaluable when the Internet is congested or you hook up to someone with a slower machine.

I tested Speak Freely with a friend locally. Without changing a thing the sound was excellent. Then my friend tried it with his mother and reported back, "The sound was horrible."

My friend's problem was that he and his mother didn't tweak it enough to get good sound quality based upon the type of Internet connection and types of machines they were using, not to mention the relative network traffic.

Most people who don't want to fool around with settings probably won't like Speak Freely, but I highly recommend it if you are the type that likes to get under the hood and fiddle with the engine. You can fiddle with compression rates, change transfer protocols, and work around problems commonly encountered with various sound cards and Winsock files. If you want to know more about the program you are running, and get a better understanding of why some Internet phone programs work better than others, then I recommend you keep Speak Freely loaded on your hard drive.

I use Speak Freely in the seminars I teach to demonstrate sound compression codecs.

TeleVox

Voxware, Inc.
305 College Road East
Princeton, NJ 08540
Phone: (609) 514-4100
Fax: (609) 514-4101
E-mail: info@voxware.com
Home Page: http://www.televox.com
Platform: Windows 3.1, Windows 95, Windows NT version 3.51

Ratings

Voice Quality: ☎ ☎ ☎ ✆
Ease of Use: ☎ ☎ ☎ ☎
Overall Product Features: ☎ ☎ ☎ ☎
Worked the First Time? Yes
Server-Based? Yes
Text Chat? Yes
File Transfer? Yes
Voicemail? No
Cost: Free, with unlimited talk time
Linkable via Web Page? Yes
Nifty Features: Ability to add sound effects to your voice

Minimum Requirements

- 486DX/66MHz PC
- 8MB RAM
- 2MB free disk space

note *The Windows 3.1 version of the Netcom Netcruiser Internet dialer is not supported by TeleVox.*

Features

A very easy-to-use interface (see Figure 6-10) offers hands-free operation for those who have full-duplex sound cards. When you first launch TeleVox,

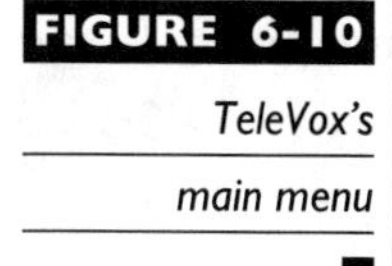

FIGURE 6-10

TeleVox's main menu

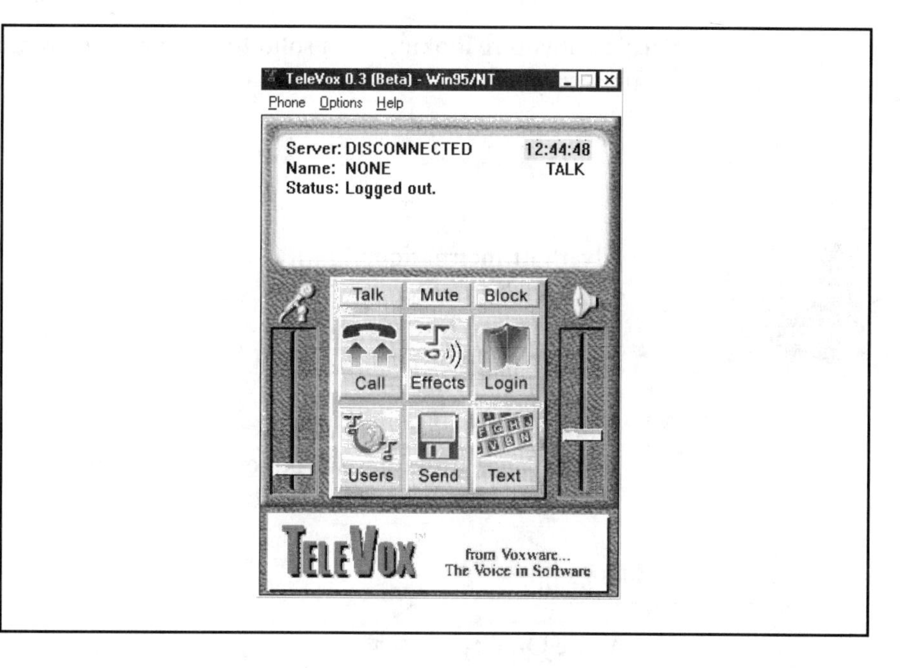

the software connects to a server, listing your name and other active TeleVox users in the directory. You can block calls, identify callers before you connect, and create private groups only you and the group can see via the server.

TeleVox provides different effects that let you change the way your voice sounds. Text chat is also available, as is the ability to transfer and receive files during conversations.

Personal Opinion

It's reliable. The servers never seem to go down and although they may be busy, never seem overloaded, and you can always find someone to talk to. Plus, it works with low-end Windows 3.1 computers.

The interface, with it's big, well-labeled buttons, makes this a program just about anybody can use, and since the standard version will always be free according to the company, it's a worth a try. The sound quality is always crystal-clear. Sometimes it will overflow your speakers, but nine times out of ten you'll get very audible sound. A little muffled, but understandable.

The only bad part of the program is the sound effects. They're funny, but they don't really make for good conversation over the busy Internet network.

Definitely try it out. I've talked to people all over the world who are using TeleVox to keep in touch with relatives, suprisingly within close proximity. It seems I've met more people in Texas using TeleVox than from any other place. So if you're looking for a solid Internet phone program, and a Texan to talk to, try TeleVox.

SoftFone

SilverSoft International, Ltd.
1108 Kashif Center
Shar-e-Faisal, Karachi Pakistan
Phone: 92 21 567-8171
Fax: 92 21 568-7393
E-mail: softfone@silver.com.pk
Home Page: http://www.pak.net/softfone.htm
Platform: Windows 3.1, Windows 95

Ratings

Voice Quality: ☎☎☎☎
Ease of Use: ☎☎☎✆
Overall Product Features: ☎☎☎✆
Worked the First Time? No, but I did get a good connection on the second try.
Server-Based? Yes, but you can use an IP address too.
Text Chat? No
File Transfer? No
Voicemail? Yes
Cost: $19.95
Linkable via Web Page? No
Nifty Features: Ability to record different greeting messages for use on the four different lines

Minimum Requirements

- 486DX/33 MHz PC
- 4MB RAM
- 2MB free disk space

Features

SoftFone (see Figure 6-11) offers multiple phone lines, full-duplex operation, a built-in answering machine, adjustable compression rates, and the ability to connect via a name server or an IP address.

The voicemail feature allows you to record your own special greeting for each line. It also shows you the number of messages received and lets you save those messages to your hard disk. You can also mute lines and put callers on hold.

The price of SoftFone is $19.95, which gets you two licenses (one for you and one for another person you plan to call). Once you buy the product, you can test it for 30 days, and return it if you're not fully satisfied.

Personal Opinion

I really liked the interface on this program: it's very clean and simple. Adjusting compression is just a matter of clicking on the modem speed button. By recording and then playing back your voice at the varying speeds you can get a feel for what setting works best for your machine.

I had relatively good conversations with this program, but I caution you to read the help file first before you try to make the program work. Not every

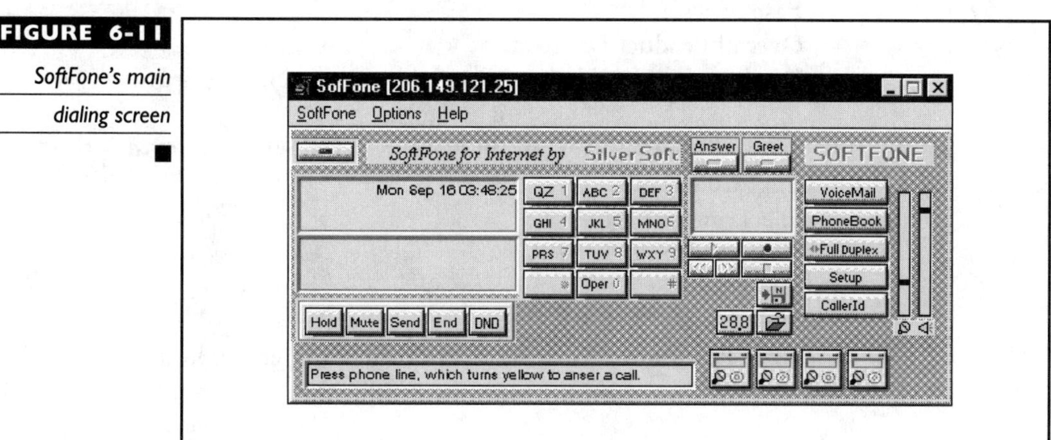

FIGURE 6-11 SoftFone's main dialing screen

button is entirely intuitive, and after fumbling around with the voicemail feature I finally read the documentation and was able to make it work.

The only reason I'm not entirely head-over-heels with the program is that connecting to the various IRC servers was a pain, since many of them didn't connect quickly or gave me error messages.

I also had a hard time finding other people who use SoftFone, as I did with IBM Connection Phone. But once I did the sound quality was relatively smooth. Again, I think this may be a sleeper of a program. If you are looking for a simple program with voicemail to communicate with one or two people, I'd definitely take SoftFone for a spin. But again, read the help file first before you delve into the program and you'll avoid a good deal of frustration.

DigiPhone

Third Planet Publishing
Phone: (800) 950-3341; (214) 828-3882
E-mail: 3pp@planeteers.com
Home Page: http://www.planeteers.com
Platform: Windows 3.1, Windows 95

Ratings

Voice Quality: ☎ ☎ ☎ ☎
Ease of Use: ☎ ☎ ☎ ☎
Overall Product Features: ☎ ☎ ☎ ✆
Worked the First Time? No. I had to call tech support to get it to work.
Server-Based? No. You can dial direct via e-mail address.
Text Chat? No
File Transfer? No
Voicemail? Yes
Cost: $70
Linkable via Web Page? Yes
Nifty Features: Ability to record different greeting messages for use on the four different lines

Minimum Requirements

- 386DX/33MHz PC
- 4MB RAM
- 5MB free disk space

Features

DigiPhone is a direct-connect program that doesn't rely on IRC servers (see Figure 6-12). It works with full- and half-duplex sound cards, and achieves high-quality sound by replacing your existing sound driver with its own driver, written specifically to work with most popular full-duplex sound cards.

You connect to other users via their e-mail addresses. With DigiPhone you don't have to go through IRC servers or special name servers. To search for other DigiPhone users, you send an e-mail to DigiPhone. You'll be sent back a list of other current DigiPhone users.

Personal Opinion

As with other programs, such as TeleVox and Web.Max.Phone, the DigiPhone interface is very simplistic: a couple of buttons; no fancy voice-mail; no fancy software knobs to adjust. So you really can't complain about how easy it is to use.

The voice quality is very good, in part because DigiPhone uses some of its own proprietary sound card software drivers. These same drivers allow relatively few breakups and delays, but can cause crashes and problems due to incompatibilities with other programs. Also DigiPhone doesn't alert you to the fact that it's replacing your drivers.

The other problem? If your ISP doesn't configure this server just so you may not be able to use the program at all. Such was the case with my Dad's provider. He never had a problem with e-mail, ftping, Web browsing, or virtually any other phone program, until DigiPhone.

As the tech support technician puts it, "Some service providers just don't configure their mail servers properly and that can cause DigiPhone to just not operate. There are some people who just won't be able to use DigiPhone."

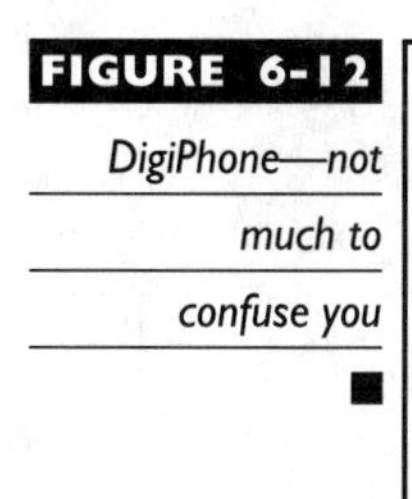

FIGURE 6-12 *DigiPhone—not much to confuse you*

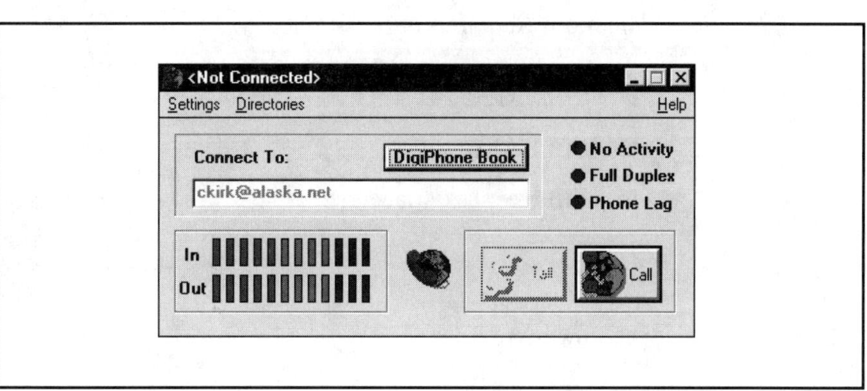

If Third Planet Publishing could fix their finicky settings, I'd say it's a good program, although compared to free programs such as TeleVox and FreeTel, $70 seems like a lot to pay. My advice? Definitely check with DigiPhone first to make sure it will work with your service provider.

CoolTalk

Insoft/Netscape Communications Corporation
E-mail: info@netscape.com
Home Page: http://www.netscape.com
Platform: Windows 95, Windows NT, Windows 3.1. A version for the Mac should be available soon.

Ratings

Voice Quality: ☎☎☎☎
Ease of Use: ☎☎☎☎
Overall Product Features: ☎☎☎☎
Worked the First Time? Yes, but hard to find people to connect to.
Server-Based? Yes, although you can call via an IP address too.
Text Chat? Yes
File Transfer? No, but does have a whiteboard feature.
Voicemail? Yes, when application is running.
Linkable via Web Page? Yes
Cost: Free in beta with Netscape Navigator 3.0
Nifty Features: Whiteboard that lets you draw, copy, and paste text and graphics interactively with the caller. Voicemail shows the little tape spinning as you record your message.

Minimum Requirements

- 486/66MHz PC
- 4MB RAM
- 5MB free disk space

or

- PowerMac

- System 7.5.3
- 8MB RAM
- 8MB free disk space

Features

CoolTalk is an Internet phone program that now comes with Netscape Navigator 3.0 (see Figure 6-13). It includes point-to-point and server-based calling, a whiteboard and text chat feature, a speed dialer, Caller ID, call screening, and a mute button.

Adjustable audio quality works on both low-speed and high-speed connections in full- or half-duplex. The CoolTalk phone book can be accessed through the CoolTalk application or via a Web page-based server list.

The answering machine kicks in when you are running the program, but is unavailable to take calls, basically offering online but not off-line voicemail such as IRIS or WebPhone. The CoolTalk watchdog watches for callers while you are on the Net, and will automatically launch CoolTalk when someone calls even if you don't have CoolTalk running.

The shared whiteboard lets you view and work with the same graphic, editing it in real time while drawing, zooming, panning, and marking it up. You can import, save, and print the file, and include TIFF, GIF, JPEG, BMP, EPS, TARGA, RASTER, and SGI files in your whiteboard.

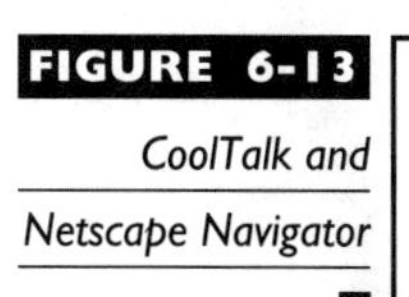

FIGURE 6-13 *CoolTalk and Netscape Navigator*

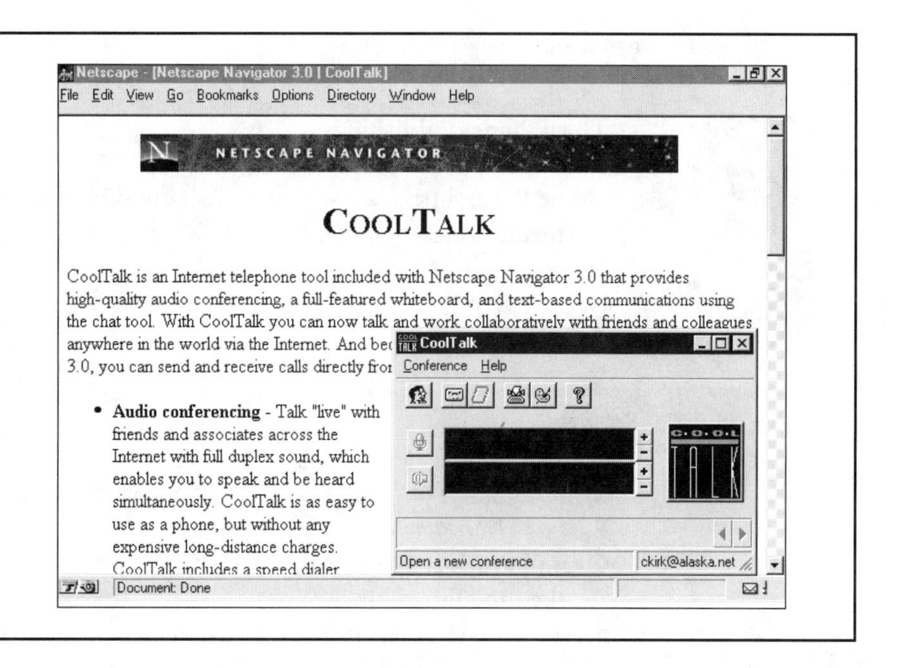

The Text Chat tool lets you type messages back and forth to the other caller, and transfer files within the same screen. You can also save the text you type back and forth into a text file. You can add callers to your address book, record your own greeting message, and include a bitmapped file of your face for other callers to see.

CoolTalk will serve as the basis for upcoming LiveMedia applications that will be based on the Real Time Protocol standard, so watch closely and see what Netscape does with this product.

Personal Opinion

The fact that you can have your all-in-one browser *and* an Internet phone program thrown in to boot is a nifty idea. And CoolTalk does deliver excellent voice quality in a relatively simple interface. However, I had a hard time finding people to talk to because of the not-so-live list of registered CoolTalk users who always seemed to be away from their computer when I called. Improving the list would make this a fantastic program.

The whiteboard feature, text chat, and file transfer options make this one of the most feature-filled Internet phone programs around. And I especially liked the ability to save my chats and whiteboard scribbles. Watching the little tape recorder spin around can be pretty mesmerizing too.

If they fix the directory listing problem I think they'll have a winner. If and when they totally integrate it into the Netscape browser, look for this to be the most widely used and popular of the Internet phone programs.

DigiPhone Mac (formerly e-Phone)

Third Planet Publishing
E-mail: support@planeteers.com
Home Page: http://www.planeteers.com/digifone/ephone.htm
Platform: Macintosh IIsi to PowerMac

Ratings

Voice Quality: ☎ ☎ ☎ ☎
Ease of Use: ☎ ☎ ☎ ☎
Overall Product Features: ☎ ☎ ☎ ☎
Worked the First Time? Yes
Server-Based? Yes, although you can call via an IP address too.

Text Chat? No
File Transfer? No
Voicemail? No
Linkable via Web Page? Yes
Cost: Under $70. Available directly from Third Planet Publishing.
Nifty Features: The e-Phone Alert, a small, background-only application listens for incoming calls and lets you know when someone is calling without having to launch the main application.

Minimum Requirements

- Macintosh IIsi (20MHz 68030)
- Sound Manager 3.*x*

remember *If you're using a PowerMac with System 7.1.2, you'll also need ObjectSupportLib in your Extensions folder.*

Features

e-Phone/DigiPhone 1.2 will run on any Macintosh from a Mac IIsi to the fastest PowerMac and can accommodate the slow machines through the use of its custom, low-overhead codec. Anything faster than a IIsi can use the standard GSM compression for relatively good full-duplex sound depending upon the model of Macintosh you are using.

note *The following models support only half-duplex sound—PowerBook 1xx, Duo 210, 230, 250, 270, 270c, 280, 280c, LCII, LCIII, IIsi; IIvx; Iivi, 605 LC475, Performa 47x, 46x, LC630, Quadra 630, and Performa 63x.*

You can connect directly to the other person if you know their IP number, or you can tell them to meet you in a "NetPub." When someone calls you, e-Phone provides caller ID, displaying their name, host name, and geographical location. You can also have multiple active calls going on at the same time, which eliminates many busy signals.

The address book lets you save other users' profiles so you can quickly call someone without having to know their IP address.

e-Phone/DigiPhone also supports the VAT protocol, which means you can talk via another program offering the same protocol, such as Maven, Speak Freely, or the video conferencing product CU-SeeMe.

Personal Opinion

e-Phone/DigiPhone was the first commercial Internet phone program I tried and really was what convinced me Internet telephony was a real thing. The sound quality is very good, provided both you and the other caller have chosen just the right sound compression for your machine and network connection *and* you have a fast Macintosh.

Although they say it works just fine on slower Macs, I don't recommend less than a low-end PowerMac or at least a Quadra. The delay is too great with slower Macs. And don't try using it with a slow Mac *and* a QuickCam. You won't hear much of anything but garbled sound.

Like Speak Freely, DigiPhone lets you adjust the compression settings, but many people don't bother. If you plan to use the program with another person, the best thing to do is experiment with the different settings to get just the right sound quality.

The interface is very easy to use, and it really only takes a few clicks of the mouse to start talking with someone else. But there's a problem. Third Planet Publishing has just recently made the Mac demo available on their Web site, so you may not find many Mac users just yet. If you want to try it, but you have no one to talk to, send me an e-mail (**netphones@aol.com**), and we can arrange a time to test it out.

ClearPhone

Engineering Consulting
583 Candlewood St.
Brea, CA 92621
Phone: (714) 671-2009
Fax: (714) 255-9984
E-mail: radiobob@earthlink.net
Home Page: http://www.kaiwan.com/~radiobob/
Platform: Newer Macintoshes

Ratings

Voice Quality: ☎☎☎☎☎
Ease of Use: ☎☎☎✆
Overall Product Features: ☎☎☎☎✆
Worked the First Time? Yes
Server-Based? Yes, although you can call via an IP address too.
Text Chat? Yes

File Transfer? Yes, via a whiteboard feature
Voicemail? Yes
Linkable via Web Page? Yes
Cost: Currently free
Nifty Features: Exceptional voice quality

Minimum Requirements

- Macintosh
- System 7.5.3
- Open Transport 1.1
- 8MB RAM
- 5MB free disk space

Features

Although it requires a high-powered Macintosh, ClearPhone delivers excellent voice quality via a patent-pending voice and time compression technology (see Figure 6-14). You can also select the size of your transmissions from 3 to 11KHz.

The program can function in the background waiting for callers, answering the phone on the first ring. Or, if you are currently speaking to someone with ClearPhone and another call comes through you'll be notified and you can put the other caller on hold and pick up the second call.

Call-waiting and hold, plus simultaneous connections, make ClearPhone work much like a standard telephone.The whiteboard feature lets you send pictures and text from your Macintosh clipboard. Working in conjunction with other programs you could copy and paste the entire whiteboard session and save it as a sort of transcript.

The "Pubs" are servers where you can meet other ClearPhone users, schedule calls, or leave messages on a message board. You can call people via the pubs or directly from an IP address.

Personal Opinion

ClearPhone is a great program, although I think the interface could be refined. The voice quality is very clear and I experienced very few, if any, gaps in my conversations.

The only problem is that it requires a hefty Mac and Open Transport, something a lot of Mac users wrestle with or simply don't have. Restricting it to only those that run Open Transport severely cuts down on the potential Macintosh user base.

FIGURE 6-14

ClearPhone

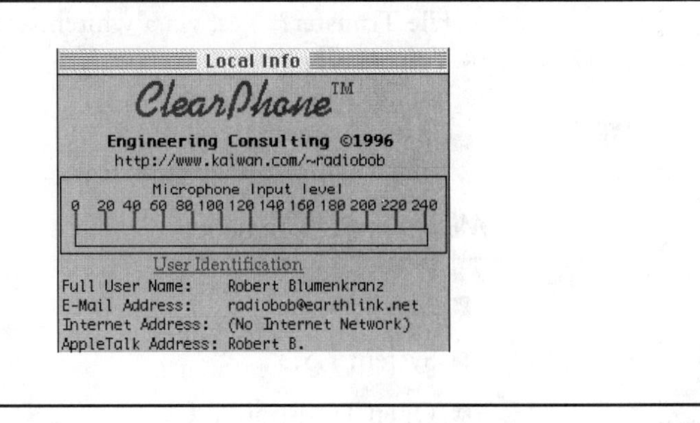

You also have to do a fair amount of configuring of both the program and your Web browser if you want to use the Pub feature of the program. Again, a little refinement on the interface to make it more Mac elegant is really needed here. In my opinion the hoops you have to jump through to make Open Transport work may turn off the novice Mac user.

The ClearPhone program is in a constant state of flux, so hopefully with every update you'll see refinement to the interface. It has promise, especially if the creator of the program ports it to the Windows 95 platform.

PGPfone

Pretty Good Privacy, Inc.
555 Twin Dolphin Dr., Suite 570
Redwood Shores, CA 94065-2102
Phone: (415) 631-1747
Fax: (415) 631-0599
Home Page: http://web.mit.edu/network/pgpfone or
http://www.pgp.com
Platform: Macintosh and IBM (usable across platforms)

Ratings

Voice Quality: ☎ ☎ ☎ ☎ (very good quality if you tweak it)
Ease of Use: ☎ ☎ ☎ ☎ ✆
Overall Product Features: ☎ ☎ ☎ ☎ ✆
Worked the First Time? Yes
Server-Based? No, IP only.
Text Chat? No
File Transfer? No
Voicemail? No

Linkable via Web Page? No
Cost: Currently free
Nifty Features: Encrypted, totally secure, voice conversations

Minimum Requirements

- 486/66MHz PC
- 4MB RAM
- 2MB free disk space

or

- 25MHz Macintosh (PowerMac recommended)
- 68LC040 processor
- System 7.1
- Thread Manager 2.0.1 and ThreadsLib 2.1.2
- Sound Manager 3.0

Features

When you dial someone's IP address with PGPfone, no one else can eavesdrop on your conversation. PGPfone takes your voice from a microphone, then digitizes, compresses, and encrypts it before sending it off to the other PGPfone user. If you wanted to protect your privacy fully, you could also use PGPfone to talk to other people over modems, thus truly protecting any conversation from eavesdroppers.

PGPfone does not need any secure channels in order to exchange cryptographic keys before the conversation begins. The two parties negotiate their keys using the Diffie-Hellman key exchange protocol, which reveals nothing useful to a wiretapper, yet allows the two parties to arrive at a common key that they can use to encrypt and decrypt their voice streams.

There is not much to PGPfone (see Figure 6-15), but it does offer the ability to adjust the microphone and speaker volume while connected, the ability to monitor packet transmissions, and the "push-to-talk" feature for half-duplex sound card users.

Personal Opinion

I must say I liked the simplicity of PGPfone. It does one thing—provide secure voice communications between Mac and/or PC users, and it does it well without a lot of bells and whistles.

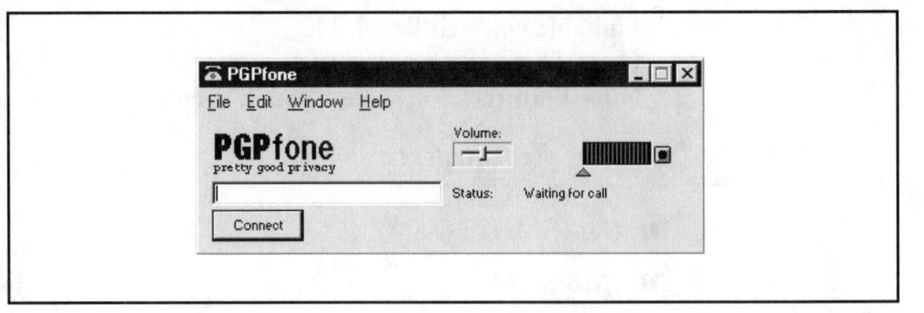

FIGURE 6-15 *PGPfone: Straightforward and not flashy*

But what may turn some people off (as it did one of my friends), is that you have to adjust the compression settings just so for clear voice over long distances. Unlike some programs that auto-adjust to the situation, PGPfone doesn't, and that means you have to really test different combinations of primary and secondary compression schemes before you get what would be considered great voice quality.

My first conversation with PGPfone was on a Mac with a friend not too far away. It was just like a telephone conversation. My second was on a Pentium with my Dad in Houston. It was awful; not because of the program, but because we had not adjusted the voice compression to match our machines and the speed of the Internet. So don't expect it to sound great the first time, but be happy if it does.

As far as the encryption, secure voice communications part, this does appear to work and work well. I just have never had anything that important to say, but if you are a business looking to protect those trade secrets, PGPfone is the only Internet phone program that provides virtually tap-proof connections.

WebTalk

Quarterdeck Corporation
13160 Mindanao Way
Marina del Rey, CA 90292-9705
Phone: (800) 354-3222
Fax: (310) 309-4217
E-mail: info@quarterdeck.com
Home Page: http://www.quarterdeck.com
Platform: Windows 3.1, Windows 95, Windows for Workgroups

Ratings

Voice Quality: ☎☎☎½
Ease of Use: ☎☎☎½
Overall Product Features: ☎☎☎☎

Worked the First Time? Yes
Server-Based? Yes
Text Chat? No
File Transfer? No
Voicemail? No
Linkable via Web Page? No
Cost: Under $50
Nifty Features: Shows you a world map pinpointing where you and the other caller are located.

Minimum Requirements

- 486DX/33MHz PC
- 8MB RAM
- 13MB free disk space

or

- Macintosh Quadra or PowerMac
- 8MB RAM
- 12MB free disk space

Features

WebTalk (see Figure 6-16) offers very good sound quality, adjustable compression settings, and cross-platform compatibility so that Mac and Windows users can talk to each other. WebTalk can work with Web browsers such as Quarterdeck's Mosaic Web browser, Netscape, or Internet Explorer, to offer automatic connection to other WebTalk callers by simply clicking on a link in a Web page. On Quarterdeck's WebTalk home page, you can sign up to be listed in the active Web-based directory, along with other active users.

Or you can use the program by itself to call other WebTalkers, either directly, with an IP address, or via Quarterdeck's server.

There are also virtual meeting places listed on the server so that you can identify and talk with people interested in specific topics. You can also use the text chat window to type information to the other caller.

Personal Opinion

The features and sound quality are very good, but the configuration and interface are somewhat awkward and will turn a lot of people off simply

FIGURE 6-16 *WebTalk by Quarterdeck*

because it's not very easy to set up. You are constantly moving between Web browser and phone program, neither of them suited to accommodate each other on the same screen. Finding people was also somewhat confusing since people came and went quicker than I could click back and forth.

But once it's running, I've got to say, seeing a world map and the link between you and the other caller graphically really puts the whole Internet phone thing, and the Internet in general, into perspective.

This graphical interface, although nothing but fluff really, is something to see, and worth the effort of plodding through the configuration. The sound isn't really any better than most, and there are easier ways to serve up a list of people than clicking back and forth between your Web browser and WebTalk's nearly full-screen interface, but it isn't a bad product at all.

Unfortunately, Quarterdeck has pulled the demo from their Web site, with no indication of when the revamped version will appear! You can still buy WebTalk from many major computer stores, but keep checking the Web site for a chance to try before you buy.

VocalTec's Internet Phone

VocalTec
35 Industrial Parkway
Northvale, NJ 07647
Phone: (201) 768-9400
Fax: (201) 768-8893
E-mail: info@vocaltec.com
Home Page: http://www.vocaltec.com
Platform: Macintosh and IBM (usable across platforms)

Ratings

Voice Quality: ☎ ☎ ☎ ☎ (very good quality if you tweak it)
Ease of Use: ☎ ☎ ☎ ☎
Overall Product Features: ☎ ☎ ☎ ☎
Worked the First Time? Yes
Server-Based? Yes, but you can connect direct.
Text Chat? Yes; also includes a whiteboard and document sharing.
File Transfer? Yes
Voicemail? Yes, you need the voicemail program.
Linkable via Web Page? Yes
Cost: $49.95
Nifty Features: Cross-platform compatibility means that Macintosh users can talk to PC users with no adjustments or special software. You'd never know you are talking to someone with another type of computer.

Minimum Requirements

- 68040 Mac (PowerPC recommended)
- 68LC040 processor
- System 7.5.1 (System 7.5.3 recommended)
- Will work with Virtual Memory or Ram Doubler
- QuickTime 2.0 or later
- Sound Manager 3.1 or later

or

- 486/66MHz PC
- 8MB RAM
- 4MB free disk space (more if you want to store lots of voicemail messages)

Features

Internet Phone is feature-packed. Not only is the program cross-platform compatible, meaning you can call your Mac friends if you have the PC version, and vice-versa, Iphone (as it is nicknamed) also has as many of the same features that most business telephone systems have. These include call holding, call-waiting, voice muting, call blocking, identification of caller so you can screen your calls, direct calling to another person by their e-mail address or nickname, plus an animated assistant (see Figure 6-17) who notifies you whenever the phone rings, when the other person is talking, or when the line you're calling is busy.

With Iphone, you can also transfer files, share documents via the white-board facility, and text chat (in case voice quality is not clear). The software supports full- and half-duplex sound cards, and its various built-in voice compression schemes can be adjusted manually. With certain microphones and sound cards, Iphone provides automatic voice activation.

Iphone hosts some of the most active chat rooms of any Internet phone program. The last time I looked, there were over 235 different rooms available. And you can link your Iphone nickname to your Web page so people can call you directly with the touch of a mouse button.

Personal Opinion

Iphone is the granddaddy of all Internet phone programs, and by far the most widely used. It's revamped interface, with its cartoon Automated Assistant, is great for novices and a big improvement over earlier versions. But the PC version still has a way to go to match the relatively clean interface of the Mac product. The PC version pops up two screens, one for Iphone itself and another for the chat rooms. Navigating back and forth between the two can be

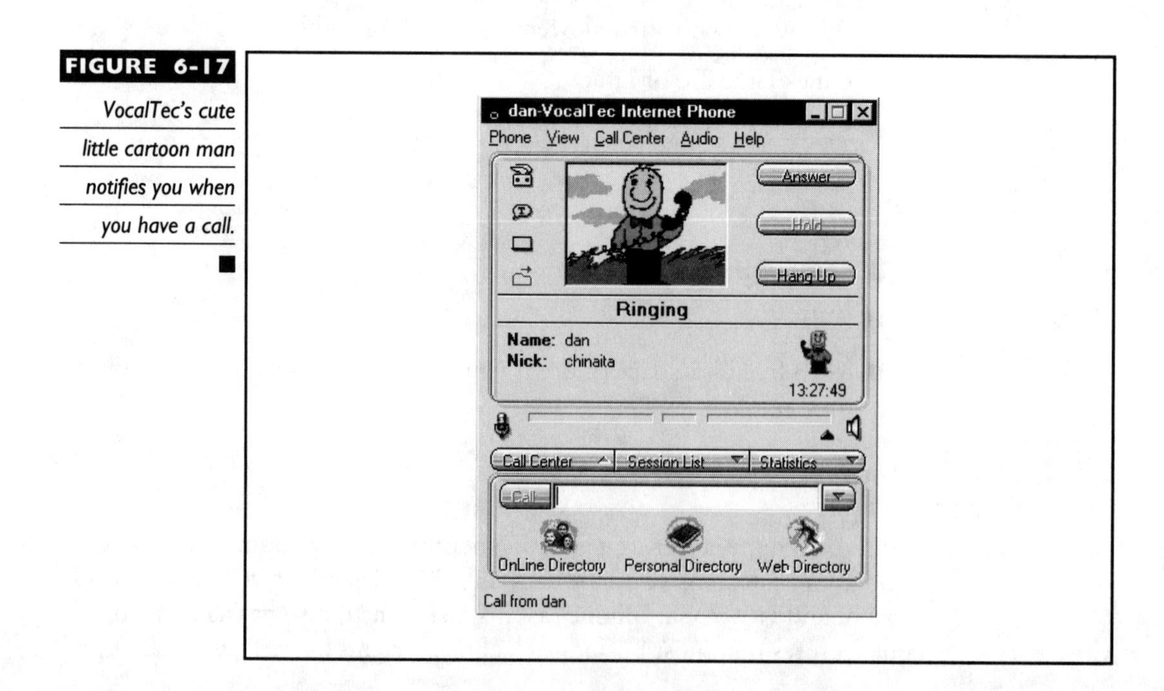

FIGURE 6-17 VocalTec's cute little cartoon man notifies you when you have a call.

a nightmare. I wish the programmers would devise some way of integrating the chat room listing feature into the actual phone program itself.

Since Iphone is so widely used, you'll have more people to talk to, especially with all those chat rooms available. But beware! With all those people comes a percentage of pervert types who like to do sex chats, so I'd be careful using this program around kids. Even some of the names of the chat rooms are, shall we say, rather graphic.

With all those people talking, it does take quite a long time to connect to the servers, and sound quality is often compromised. While I have had some great conversations on Iphone, I've also been thwarted by some totally inaudible ones. Unlike Intel Phone, where a very clear connection can almost always be guaranteed, it's really the luck of the draw with Intel Phone. I must say I got far better connections with the Mac version than the PC version.

I certainly would put this on the list of products to try, if only to see the little cartoon guy move about the screen. If you have a Mac, I definitely recommend you try it, although you may need to play with the different sound compression schemes to get a perfect connection. Finally, if you're looking for a new friend from a different state or country, you're bound to find them in one of the Iphone chat rooms.

Interphone

E-mail: support@interphone.com
Home Page: http://www.interphone.com
Platform: IBM

Ratings

Voice Quality: ☎ ☎
Ease of Use: ☎ ☎ ☎ ☎
Overall Product Features: ☎ ☎ ☎
Worked the First Time? No, couldn't connect, but finally did the third time.
Server-Based? Yes, but you can connect direct through an e-mail address or nickname.
Text Chat? No
File Transfer? No
Voicemail? No, but you can leave a text message.
Linkable via Web Page? No
Cost: Currently free
Nifty Features: Ability to use a nickname instead of e-mail address or IP address, making it easier for people to call you.

Minimum Requirements

- 486/33MHz PC
- 8MB RAM
- 2MB free disk space

Features

Interphone (see Figure 6-18) offers direct connections to other callers through their servers. You can call up the other person via their e-mail address or the nickname they have saved in their configuration. You can also save people's addresses through the autodial feature, and then take advantage of one-button dialing.

You can also leave text messages, put callers on hold, and have multiple conversations on different phone lines. To find other users, simply click the Phone Book button; callers are listed under their state or country.

Personal Opinion

I REALLY liked the interface for Interphone, but I can't say much about the sound quality. I tried calling Dad with the program and had a hard time connecting the first couple of times. When we did finally connect, his voice sounded so much like Charlie Brown's teacher that I couldn't stop laughing.

FIGURE 6-18 *Interphone's 3-D interface*

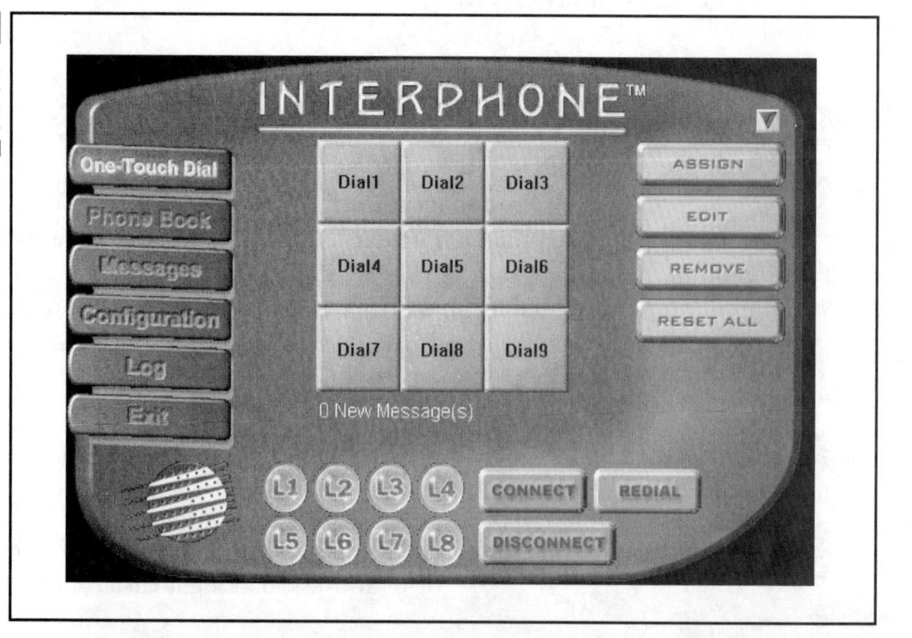

He said my voice came through pretty much audible, but with some choppiness. We tried both the 8-bit and 16-bit settings, with both normal and high compression, but none of these changes seemed to affect the sound of his voice on my end. Perhaps I missed something, but there doesn't seem to be much documentation or information on the Interphone Web page, so I didn't have much to go on. The Web site says a new version is in the works. The current interface is quite clean, and this could turn out to be a very good program once its creators fix whatever problem caused such poor sound quality on my end.

One downfall of this product is its low number of users. Most states and countries appearing in the directory list a fair number of users, but the four times I tried the program I could only find one person online. Still, I do recommend giving Interphone a spin, especially if you know of someone else who's willing to try it out.

Internet Multimedia

Hani Abu Rahmeh, author
E-mail: hani@hyperion.demon.co.uk,
Home Page: http://ourworld.compuserve.com/homepages /hani/imm.htm
Platform: Window 95, Windows 3.1

Ratings

Voice Quality: ☎☎☎☎
Ease of Use: ☎☎☎✆
Overall Product Features: ☎☎☎☎
Worked the First Time? Yes, but I had problems connecting via IRC server on second call.
Server-Based? No. You can call direct or use the IRC servers.
Text Chat? Yes
File Transfer? No
Voicemail? Yes
Cost: $15
Linkable via Web Page? No
Nifty Features: The answering machine lets people leave messages on your computer, as long as you are connected to the Internet and running Internet Multimedia in answering machine mode.

Minimum Requirements

- 486DX2/66MHz PC
- 8MB RAM
- 2MB free disk space

note *If you don't have a sound card and speakers, you can still use the chat and whiteboard features offered by Internet Multimedia.*

Features

Internet Multimedia is a relatively easy-to-use Internet phone program that allows two people to talk over the Internet, leave voice mail messages, use a common whiteboard, and text chat. You can look up other logged-in users in the IRC chat server listing, and simply click on the name of the person you want to talk to. If the other caller is running Internet Multimedia, but is away from the computer, you can leave a voice mail message.

The main screen (see Figure 6-19) offers one-button access to all the program's features. When a button is clicked, the corresponding feature is displayed in a separate window. An address book feature lets you store frequently called IP addresses. You can also add to the list of IRC chat servers and toggle between normal and high voice quality.

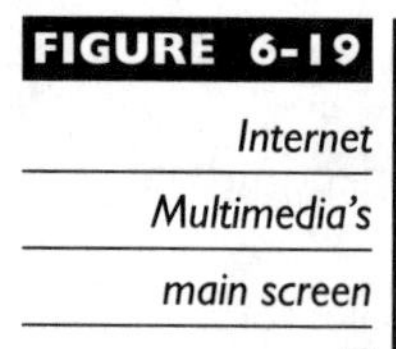

FIGURE 6-19 *Internet Multimedia's main screen*

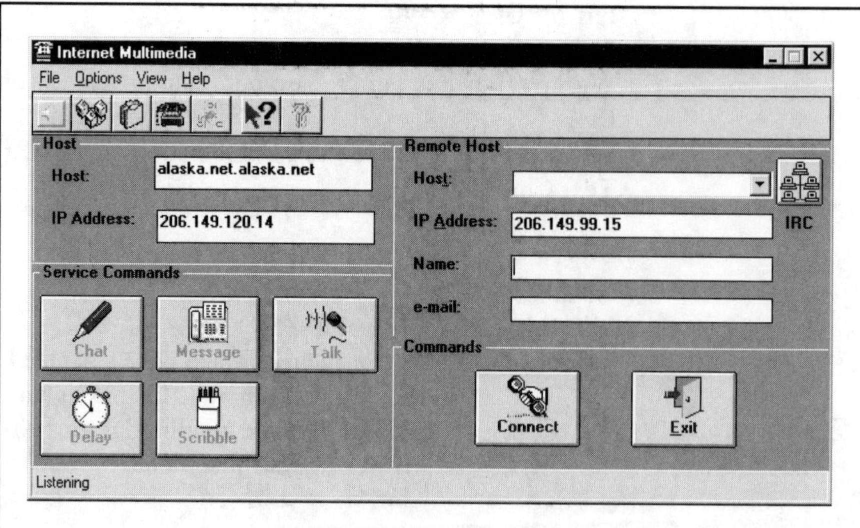

You can click the Delay button to find out how long it takes for your voice to reach the other person's machine. Clicking Help button lets you access help with any of the program's features.

Personal Opinion

The sound quality was very good, and I had no problem connecting the first time directly via IP address. I didn't have much luck through the IRC servers, however. But that could have been attributed to the people I was calling; they may not have been on the right servers when I called.

Internet Multimedia is a pretty simple, straightforward program if you're making a call, but you have to physically select "server mode" to get it to answer incoming calls. The same is true of the answering machine. Unlike similar features in other Internet phone programs, this answering machine doesn't automatically kick in when you don't answer the phone. You must first set the program to "server mode," which simply instructs it to answer incoming calls and take messages.

The IRC servers never seemed to list anybody using the program, and I suspect that not too many people are currently using it. That's too bad, because it does have some nice features, such as the answering machine, relatively fast text chatting option, and somewhat muffled—but otherwise clear—audio.

The only thing I didn't particularly like about the program was the whiteboard. It popped up in another window and was very klunky to use, not at all like other whiteboards that feature standard paint-type palettes.

The author has plans to add videoconferencing, file transfer capability, document sharing, and full-duplex speech in the future. The version I tested still ran in half-duplex mode, but it let me tweak the delay so that the conversation wasn't too slow. If the author adds the promised enhancements, then I think Internet Multimedia, like so many of the other free and shareware-type Internet phones, will rival many of the commercial products on the market.

TELEPHONE

Chapter 7

Putting the Internet Phone Programs to Work

COMPUTERS should be easier to use.

My Dad's adventures in the high-tech world illustrate the fact that computers still have a long way to go before they are as easy to use as the phone. He has no problem writing a letter, checking his e-mail, or cruising the Web—in fact, my trusted beta tester is the typical average Joe computer user. (Of course, it helps that Dad's name *is* Joe.) But when it comes to downloading and decompressing files, or running esoteric installation programs, Dad often stumbles. As Mom will attest, sometimes he gets downright frustrated and gives up. Frankly I can't blame him. It seems every Internet phone program has its own installation idiosyncrasies and its own unique configuration. And the current assortment of Internet phones offers a wide range of techniques to get users connected and talking.

In this chapter I'll shed some light on how to download, install, and configure a variety of Internet phone products. Because Internet telephony is such a new and ever-changing breed of technology, the best advice I can give you up front is to be patient—and *make sure* you've backed up your system before you start. Not everything will work the first time. Nor will all of these programs seem to operate logically—at least not at first. But once you master a few simple steps, you'll be well on your way to using your computer for something really exciting: talking to people all over the world, and best of all, doing it FOR FREE!

First, Back Everything Up!

Before we jump into the actual installation process, stop right now and back up your system. No whining or complaining, just do it. Remember, anytime you install a new piece of software it's a good idea to have a backup of your hard disk, or at least of all the files you couldn't bear to lose if something went wrong. This is particularly important with Internet phone programs, since they oftentimes change the system information files that make your computer work.

remember *Programs such as FreeTel, VDOPhone, and WebPhone make changes to your system files—the Win.ini and Registry database files, for instance. Anytime a program changes something in your system there is always a possibility that device conflicts and other problems will occur. So play it safe and back up your important files.*

If you don't have a tape backup, seriously consider spending the couple of hundred dollars it costs to buy one. If this sounds expensive, consider the cost of the time it would take to recreate your system's current setup—and imagine your frustration and aggravation if something should go wrong. Tape offers a quick way to back up your entire system, exactly as configured, or to back up individual files of your choosing. Most tape devices connect either through the parallel port or through a special interface card that you install inside the computer.

You might also consider one of those new "Zip" or removable cartridge drives that store data on small, floppy-like cartridges. Iomega and Syquest are two of many manufacturers that make them. I recently purchased a Syquest EZ-135, and I simply love it. It connects to the parallel port in back of my PC, and has an additional connector so I can hook it up to my system without having to disconnect my printer. The cartridges used by these devices are small enough to fit in a shirt pocket. Each cartridge holds up to 135 megabytes of data, and the drive itself is fast enough that I can use the cartridge as I would a second hard disk.

I've got almost the same thing on my Macintosh. A couple of years ago I bought an Iomega Bernoulli 230 drive, and I love it too. I never run out of space on my Mac, because I can always back up anything I need onto a 230-megabyte cartridge, or simply use the cartridge as an extra hard disk. Now I have unlimited disk space on both machines!

tip *Look for the storage capability of these cartridge drives to increase dramatically. Already several vendors are making gigabyte cartridge drives. Wow!*

The main difference between a tape backup device and a removable cartridge drive is that tape stores your data in a fixed, sequential order, while a cartridge stores it randomly, much like a computer disk. Because a cartridge drive offers random access to your data, you can use it much as you would a real disk drive. The same is just not true for a tape device. Retrieving a file from tape storage involves winding and/or rewinding the tape until you reach the file's location. This can take time, especially if the file you want is buried on the tape behind lots of other data. Tape is fine for archiving your files, but it's simply not practical for storing data that you plan to use regularly.

The main (and almost only) advantage of tape storage is that many tape storage systems come with special software that will back up your files automatically at a specified time. This feature makes it easy to back up data regularly, without much effort. Tape storage is great for businesses that archive lots of data; automatic daily backups can virtually ensure the safe storage of important company information. An individual like me, however, usually doesn't need to back up massive quantities of data. I just need enough space to store my individual files—and once they're backed up safely, being able to access them easily is a real plus.

If you have Windows 95, make sure you either use the Backup program to back up your entire system or copy the entire contents of your drive to a removable cartridge drive (or if you don't have one of these, to floppy disks). If you don't have time to do a complete backup, at least back up your Win.ini and System.ini files, and run the Registry Editor (RegEdit) from the Start/Run menu to "Export" your Registry database, making sure you make copies to either tape, cartridge or diskette. If you're a Windows 3.1 user, make sure you back up at least your Autoexec.bat, Config.sys, Win.ini, and System.ini files. If you don't have a tape or cartridge drive, these few files together shouldn't take up more than a single disk.

for mac users *Back up your entire System folder, or at the very least back up your System, Finder, Control Panels, and Extensions folders, plus any System Enabler files. (These will probably take you several diskettes if you do your backups the old-fashioned way.)*

Prepare Your Computer

The best way to ensure a smooth installation is to make sure your computer is running properly, has enough hard disk space to store your phone software and all the program's associated files, and is actually capable of running the program. If you have intermittent problems with your computer, or if it's barely capable of running the programs you have now, then you're likely to have trouble running any Internet phone program.

Your connection to the Internet is just as important as your computer. If you have a slow connection, or have problems connecting or staying online for long periods of time, or if your system hangs, crashes, or freezes when you're on the Net, then contact your service provider to find out if they're experiencing any problems, or if they've upgraded their system and possibly changed any network numbers. Oftentimes upgrades and system expansions will change the network numbers you might have originally used to log in to the Internet. If your computer is looking for a server at a certain address, and your setup is pointing to the wrong server, it may take longer to find the right computer at the right location, or your machine may simply refuse to connect you.

Even if this isn't the case, make a point of asking your provider for advice periodically. For example, find out whether they recommend using different connection software than the one you are currently using, or if they have any tips on how to make your system cruise faster. Many service providers continually upgrade and update their connections and servers, so reading their technical bulletins is an excellent idea.

Start with This Checklist

Once you know you have a fairly reliable connection to the Internet, you should focus your attention on your computer. Take a good look at the following checklist. If you can answer yes to these questions, then you're almost ready to install your first Internet phone program. If not, the best thing to do is fix whatever it is in your system that doesn't measure up; then you'll be ready to rock and roll.

- Is your machine at least a 486/33MHz PC, or a PowerMac or Quadra, with a minimum 8 megabytes of memory? (This, by the way, is the absolute minimum configuration. If you simply can't upgrade to a Pentium, more RAM memory will speed up your computer considerably. But be forewarned! Some programs won't work unless you have a Pentium.)
- Does your computer have at least 30 megabytes of free hard disk space?
- Have you run ScanDisk, Norton Utilities, PC Tools, or another good disk utility to confirm the health of your hard disk? Did it run without problems? If the program did note problems, did you instruct it to fix whatever was wrong?

for mac users *At the very least, rebuild your desktop by holding down the COMMAND and OPTION keys as you restart your computer. Wait until the screen says, "Are you sure you want to rebuild the desktop?" before letting go of these keys. Rebuilding the desktop will clean up your Macintosh by getting rid of icons of files you have thrown away, and realigning files that might be slightly askew.*

- Have you defragmented your hard drive lately?
- Do your sound card and microphone operate properly? Can you record something, and then play it back, without any problems? Are you happy with the sound level and the quality of sound?
- Do you have WinZip for the PC or Stuffit for the Mac? These programs decompress compressed files. Some of the Internet phone programs stored on the *Internet Phone Connection* CD, and on various Web sites, are compressed due to their size. You'll need to decompress them before you can use them.
- Is your connection to the Internet fairly reliable? (Chapter 4 lists utilities and Web pages for checking packet loss, etc.) If your connection is unstable, or the system relatively busy, consider waiting a while for the traffic to die down. You'll get better reception.

Ready, Set, Make Folders, and Download

Once you're ready to begin, start by making a "temporary" folder and placing it on your desktop or in the root directory of your computer. Call it something like TPhones, for "temporary phones." Don't name it "files," "windows," "Iphone," or any other name that might be similar to names used by other programs. Choose a unique name so you won't mix this folder up with the actual software folder; otherwise you may not be able to easily find it later.

Why a New Folder?

Your TPhones folder will be a temporary holding space for all the compressed files you download or copy to your computer. It WON'T hold the actual Internet phone programs, rather just the installation and setup files. When you are finished completely installing the various Internet phone programs, you'll be able to safely clean out this folder knowing you won't be throwing away valuable files that make your computer run.

note *Compressed files end with the .zip or .exe extension on the PC, or the .bin or .sea extension on the Mac. These files are created when a software manufacturer uses a compression program, like WinZip on the PC or Stuffit on the Mac, to compress the actual program files and combine them into a single, smaller file.*

When you store all your compressed files and installation programs in one folder you won't have files flying all over the place—some here, some there, some not even recognizable as phone software files. And with all your phone files in one place, it will be easier to run down the list and determine which ones you should get rid of and which ones you should keep.

Here's a quick rundown of how a normal installation works:

1. First you will decompress the main file for the program you want to install. (See Appendix A for detailed instructions.) Each compressed file turns into a collection of files that makes up a single Internet phone program or installation program.
2. You'll need to run the main setup program for the software you're installing. (If you're installing one of the programs from the CD, you'll see detailed onscreen instructions once the program is decompressed, and there's also a handy interactive Help program in case you get stuck. Again, see Appendix A for more details.) The setup program will usually create a separate folder in your Windows or program directory, and put the actual Internet phone program files there, along with all the associated Help and Readme files. Most configuration files for a particular Internet phone program will be stored in its own program directory, or in a combination of your Windows, program files, and root directory.
3. Once you've set up your program, you'll need to delete the temporary files from your temporary folder. (You don't need them any longer, and they're taking up hard disk space. More importantly, you'll want to get rid of any files—Install.bat, for example—whose names might be duplicated in the *next* phone program you want to try out.)
4. From the newly created program folder, you will actually run the Internet phone program for the first time, and configure it to work with your particular system.

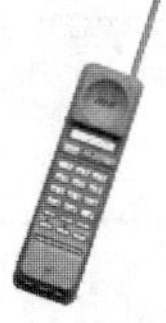

tip *Just to stay organized, you should make another folder in the root directory of your computer and name it Phones. This will be the main folder that holds all the individual folders for your installed Internet phone programs. Once you install a phone program, you should move its program folder into the main Phones folder. With all your phone programs grouped together this way, it will be easy to run down the list of folders and select the one for the program you wish to use.*

Downloading Files From the Internet

Almost all the Internet phone programs your little heart could ever desire are on the CD that accompanies this book. However, in this section I'll show you how to download additional files from the Internet. Knowing how to do this will be useful when you need to upgrade various phone products.

note *Throughout the examples I'll be using Netscape, but these steps work in just the same way with Microsoft's Internet Explorer.*

1. Connect to the Internet and fire up your Microsoft or Netscape browser.
2. Point your browser to **http://www.freetel.com**. FreeTel is a relatively simple Windows-based product and a good program to start with, mainly because you can get it up and running in just a few minutes. The current version (1.0) resides in a *self-extracting file*. This means that all the installation, Readme and Help files are wrapped up in a single file that automatically expands when you double-click its name. Self-extracting files are nice because you can extract all the files they contain without having to fiddle with compression programs such as PKUNZIP or WinZip to decompress and extract all the files they contain.

remember *You can tell whether or not a file is self-extracting by looking at its three-letter file extension. Files that end in .exe (for the PC) or .sea (for the Mac) are self-extracting files; they require no special software to unextract. But files that end in .zip (for the PC) or .sit (for the Mac) require that you use a decompressing program like Stuffit or WinZip to extract them from their compressed electronic package.*

3. Take a look at the FreeTel home page. Move your pointer over the hypertext link labeled "Download," and then look at the status bar

at the bottom of your browser's screen. There you will see the name of the file. Notice here that the file is called ft100.exe:

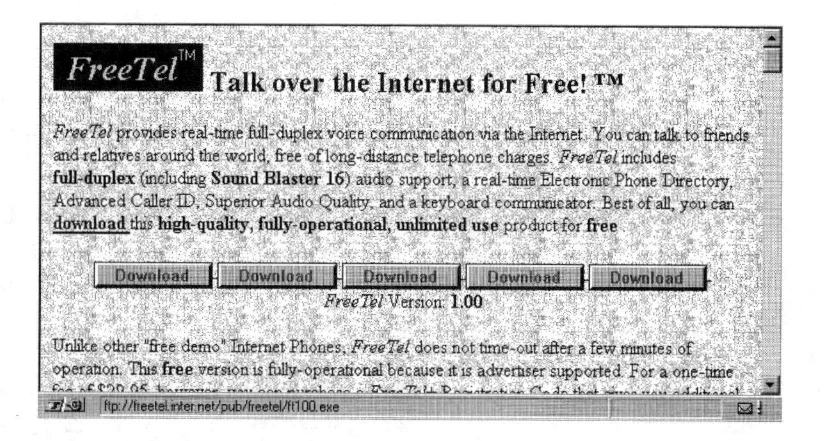

The .exe extension alerts you to the fact that this is a self-extracting file. You won't need anything else to install this program once you've downloaded it to your hard disk. Go ahead and click Download. The Save As dialog box pops up.

note *Notice that this particular hypertext link, Download, starts with the letters FTP, which stand for* File Transfer Protocol. *This means that you will immediately start to download the files the minute you click on the link, instead of being taken to yet another Web page. When you FTP a file, you are connecting to another type of server. Instead of serving up Web pages, this server's purpose involves storing files and letting you transfer them to your computer.*

At this point, the computer is asking you where you want to save the file. If you were to just click the Save button, the file would be saved in the same directory as your Netscape or Microsoft browser. But since you've set up a specific folder (TPhones) for your Internet phone downloads, you should save it there.

tip *I can't tell you how many times Dad and other novices have accidentally downloaded and saved files into their Netscape Program folder. Then Dad would say, "I thought I downloaded it, but I can't seem to find it. Maybe I should download it again." Forgetting where you placed a file can be maddening. If you're ever in this situation, take advantage of the Find option on the Start menu in Windows 95.*

for mac users *To locate a misplaced file, try using the Find option on the File menu of the Finder.*

4. To get to your temporary TPhones folder from within the Save As dialog box, click the drop-down arrow to the right of the Save In field, and scroll through all the folders that appear until you reach the Desktop icon.
5. Click on the Desktop icon. You'll see all the files and folders currently on your desktop. See the TPhones folder? Double-click it to open up the folder. You should now see the TPhones folder name in the Save In field:

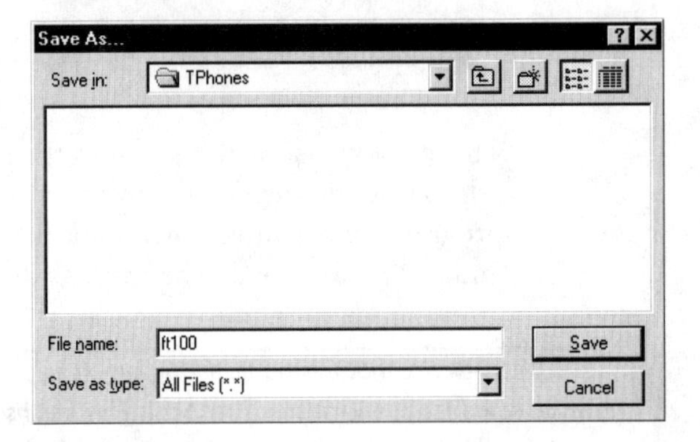

6. Simply click the Save button. (And here's a warning: don't change the name of the file! The decompression instructions embedded in some self-extracting files look for a specific filename to decompress. If that filename doesn't exist, the decompression process will stop dead in its tracks.)

   A separate window will pop up displaying the name of the file and a thermometer bar that slowly fills as the file is transferring:

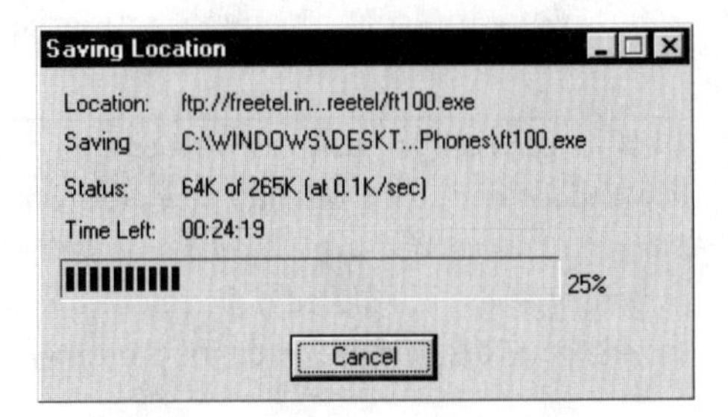

This shows you how fast your computer is receiving the file. When the thermometer has inched itself to the far right side, the file should be finished transferring from the remote file server to your computer.

note *Transfer rates above 2.5 Kbps are considered pretty fast for modem users, so if you get 2.5 or above, you have a fast Internet connection. This is excellent news, because it means you'll have relatively good Internet phone conversations.*

Congratulations. You're halfway there. Now you're ready to install FreeTel on your computer!

Installing a Phone Program

You can install almost all Internet phone programs while other programs are running on your computer, but it's not the best idea. Some programs will want to restart Windows, so if you are writing the great American novel and haven't saved it, you could lose your most recent work. I recommend that you exit from *all* application programs, including Netscape, when you're ready to extract and run *any* installation program. Now that you know that, let's get started.

note *Each phone program's setup procedure will vary slightly. Some will use the familiar Windows 95 Setup Wizard, while others use their own installation programs, as in the following FreeTel example.*

1. Once you've closed all your programs, take a look at your desktop. Locate the TPhones folder and open it up. Inside you should see the compressed FreeTel program you've just downloaded, as shown here:

2. Double-click the ft100 application icon. You should see an MS-DOS window pop up and filenames start to flash by. It takes less than a minute for FreeTel to extract itself and place the setup files in the proper directory. Notice that you didn't have a chance to tell the program *where* to extract its files. It simply extracts the files in the same folder where the compressed file is stored (in this case, your TPhones folder).

3. You'll know when the program has finished extracting all its files when the title of the MS-DOS window has changed to FINISHED-ft100. When this happens, close the MS-DOS window and take a look at your TPhones folder. You should see a whole bunch of new files: some text documents, some Help files, and others with the .dll extension. What you are looking for in this hodgepodge is a single application file entitled Setup.exe.

4. Double-click the Setup file, and a new screen will appear:

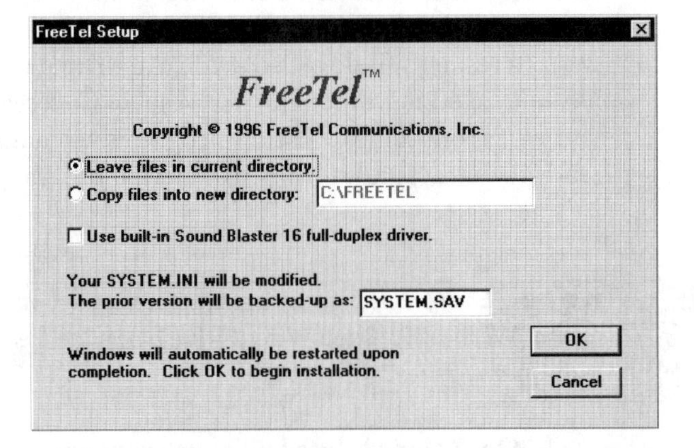

5. Notice that you have the option of copying the actual program folder into the parent directory of your choice. You should specify the Phones folder we set up earlier. (Remember, this folder will hold all of our *installed* Internet phone programs.) To do this, find the "Copy files into new directory" field, and change the C:\FreeTel setting to **C:\Windows\Desktop\Phones**.

6. Click OK. Now just sit back and let the installation program do its thing. It will create a separate FreeTel folder inside your main Phones folder, then it will install FreeTel in its new program folder and configure the software according to the settings you've designated.

remember *Many setup programs will ask you to provide your e-mail address, and to specify where your browser software is located. The programs need this information in order to connect you to the various Internet phone directory listing servers and in some cases so the program knows where to send your voicemail. Be sure to supply this information correctly; otherwise you may find that you can't connect to other people or retrieve your voicemail messages. If necessary, check your Eudora, Pegasus, or Netscape configuration for your POP, SMTP, and e-mail addresses. If you don't find them there, contact your service provider. Don't just leave it out.*

7. Once the installation is finished you should see a message indicating that the program has been installed successfully. The final message in the FreeTel installation procedure prompts you to restart Windows. Do so now, and then reconnect to the Internet.

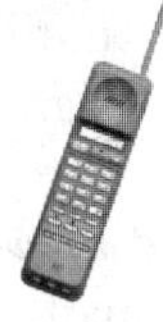

tip *Even if the program doesn't prompt you to restart your computer, you should. Oftentimes installation programs add new instructions to your Windows system files, and these instructions aren't implemented until Windows is restarted. It's a good idea to reboot your machine after installing any new software, whether or not the program is Internet phone-related.*

8. Find the Phones program folder. Inside it you should see the FreeTel folder. Now you're ready to make your first call. Exciting, isn't it?

remember *Once you've run the Setup program for your new phone software, be sure to delete the compressed file and the miscellaneous setup files from your TPhones folder. You won't have to worry about deleting files used by the program itself, because you've already installed these in the program's own folder.*

How Internet Phone Programs Work: A Brief Overview

Before you start ringing up all your friends, you should know a little bit about the way phone programs work. Internet phones bring users together in a variety of ways: directly, via an Internet phone company's server, or through a name server.

Direct Connections

The direct connection concept is pretty simple, and very similar to the process by which you make a regular telephone call. First you need the IP address of the computer you plan to call. An IP address is made up of four numbers, for example, 204.17.139.66. Each computer currently connected to the Internet has one of these unique numbers assigned to it. The numbers are used to identify all the different computers and to keep information flowing to the right destinations.

IP addresses are different than e-mail addresses. An e-mail address is a location where your mail is received. Just as in real life, you don't have to be home to get mail. In the cyberworld, your mail is dumped in an electronic holding box designated specifically for you. Only when you log on and tell the postmaster to transfer your mail to the real mailbox located on your computer is the mail actually delivered there. So when you're on vacation and you forget to stop incoming mail, your messages can sit there until they fill up your box. Only when you log on and your computer is assigned an IP address does the postmaster on your e-mail server know where to send the mail. So an IP address is the location of your computer, and an e-mail address is like a post office box where you go and pick up your mail.

Recall that some Internet providers assign dynamic IP addresses, reassigning you a number every time you call. Others provide static IP addresses that never change. Static IP numbers usually cost more, since there is a little more setup and maintenance involved. If your account is like most, you probably have a dynamic IP address. Normally, if someone wanting to use a direct-connect Internet phone program has a dynamic IP address, they just send an e-mail message to the other person to say "OK, I'm now logged on and at 204.17.139.66."

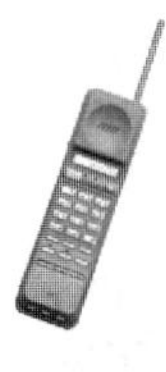

tip *To find your IP address under Windows 95, choose Run from the Start menu, then type* ***winipcfg****. This will pop up the IP configuration dialog box:*

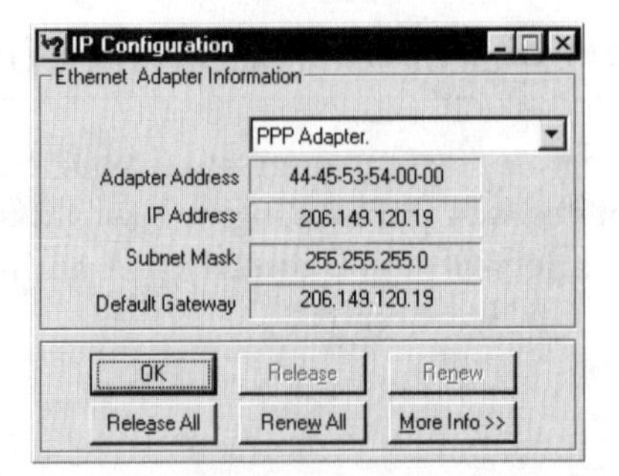

for mac users *You can find your IP address in Config PPP by clicking the Stats button. Look for the words "IP addr Local". The number that follows is your IP address.*

PGPfone and DigiPhone are excellent examples of direct-connection Internet phones. PGPfone lets you dial up anyone with an IP address, whether they have a Mac or PC. All you need to do is fill in the other person's IP address.

Once PGPfone has found the actual computer you've specified, it reads the text name that the person has given their computer, and displays that name on the connection screen. Most Internet phone programs will "resolve" IP addresses to actual computer names, thus making it easier to identify just who it is you are talking to.

Direct-connection Internet phones have one advantage over all other types: You don't have to mess with logging in to servers, and your speech won't be compromised by a slow server that is busy keeping track of all the other logged-on users. In the case of PGPfone, you can encrypt or secretly code your messages and then send them to the other party. To use encryption, you both decide on an encryption key and set up the software to transmit encrypted voice to each other. The other person must have your secret code to unencrypt the sound file. This means that your voice is secure from hackers who might want to eavesdrop on your conversation.

The only hassle with direct connect Internet phone programs? There is no indication that the other party is online. You can't check a server to see all the people currently online, and that means you can't find people all over the world to talk about the weather with. Plus, if you wanted to check out the latest and greatest direct phone program, you'd have to tell your friends about it and they would have to download it so you could try it out. Eventually, of course, your friends will get tired of being used as human guinea pigs, and you'll have no one to talk to.

Some products, such as WebPhone, offer point-to-point calling via your e-mail address, making it much easier to find your long-lost love. It's still a direct connection, but you don't have to search a server for the name of the person you are looking for, and you don't have to swap IP addresses either. This type of direct connection, which uses a server to resolve the e-mail address to the IP address of the machine you are calling, is probably the best type of Internet phone program calling interface.

I'd highly recommend direct-connected Internet phone programs like PGPfone, Intel Phone, or WebPhone if you want down and dirty, simple Internet phone connections to talk to Mom and Dad in Idaho. You connect faster, and you don't have to worry about others constantly calling you when you're waiting for the folks to call.

Name Servers

Since most people don't have static IP addresses, many companies, such as Intel and NetSpeak, have figured out a way to help dynamic IP users find one another. A *name server,* also called an *address resolution server,* lets you find another person's IP address without having to send an e-mail message or run an esoteric configuration program.

Name servers require that you first register with one of the numerous White Page directory services, such as Four11, The Internet Address Finder, Switchboard, or WhoWhere. You usually will be prompted to include your name, e-mail address, and the type of Internet phone program you are using, as shown in Figure 7-1.

By registering your name with one of these services, you're not only making it easier for other people to find you, you are also creating a spot where you'll be able to automatically link your connected IP to a Web page.

Name server-based programs work like this: Every time you run your Internet phone program, it first registers your current IP address with the server. The name server then matches your current IP address to the listing you've made. Now let's say someone finds you in the directory, and clicks on the link indicating that you are online and ready to receive a call from another Intel phone user. That person's Intel Phone program then gets your current IP address from the name server and connects to your IP address.

You still are essentially getting a direct connection, as with Intel Phone, but now you can actually see who is currently logged on to the Net and using that particular Internet phone program. This means you don't have to bug

FIGURE 7-1 *Registering with Four11, one of several directory services*

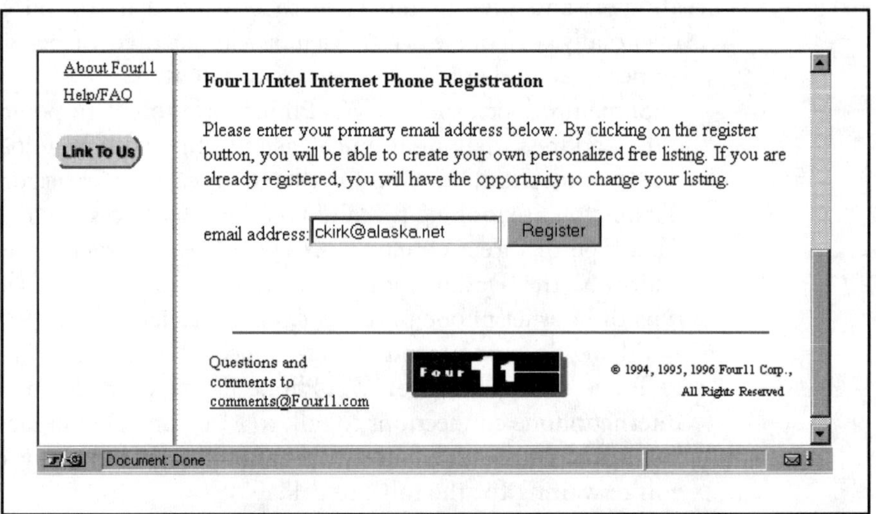

your friends to try out the latest and greatest name server-based Internet phone program. And normally with this type of connection, the sound quality is very, very good.

True Server-Based Internet Phones

Another type of Internet phone lets an Internet phone user connect to other users through a centralized server. FreeTel, TeleVox, and VocalTec's Internet Phone are examples of server-based programs.

When a user fires up her server-based Internet phone program, it automatically connects to a preassigned server. In the case of VocalTec's Internet Phone, the program connects through the Internet Relay Chat system. Other programs, such as FreeTel, use their own servers, and don't route through the Internet Relay Chat system.

Every time you connect to a server, a list of other users currently connected to that server appears. You simply click or double-click on the name of the person you want to call, and the connection is established, as shown in Figure 7-2.

Oftentimes, programs that rely on servers can be slow when you first start them up, since they have to connect to the server in order to operate. If the server is busy, it may take up to a minute for the program to connect and display all the active users. And the voice quality you get with server-based chat programs varies with each program. When servers are busy and slow, voice quality tends to degrade—sometimes quite sharply.

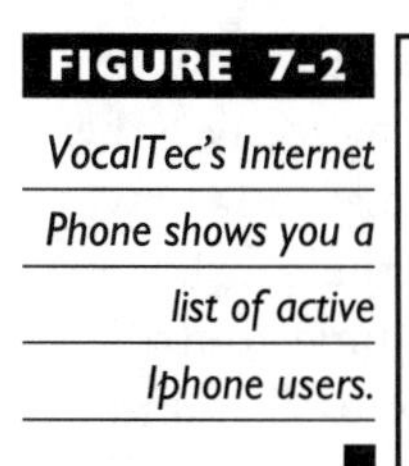

FIGURE 7-2

VocalTec's Internet Phone shows you a list of active Iphone users.

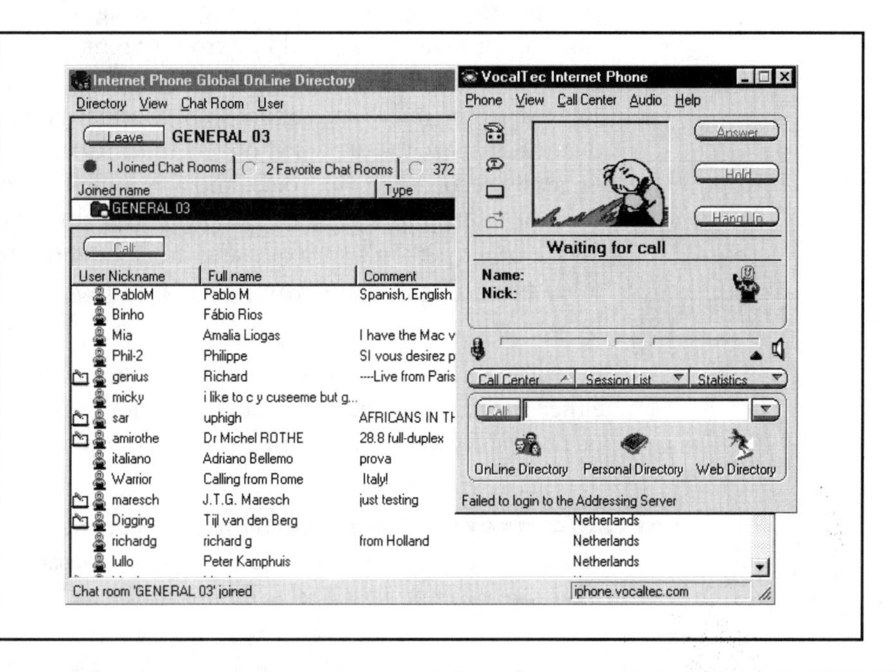

Even so, the big advantage of using a server-based Internet phone is that you don't have to know someone's IP address, or really know much more than how to double-click, to start talking. Plus, many server-based programs provide additional features, such as whiteboards, text chat areas, and the ability to share other applications.

Before You Try to Phone Home...or Wherever...

Now that you know the differences between the different types of Internet phones, you're ready to make your first call. But wait...maybe not. Before you rush in and start making random calls, make sure you are prepared. Here's a handy checklist to run through before you start dialing for dates:

1. Connect to the Internet.
2. Check your speakers and microphone to make sure they are on and adjusted to the right volume level. (I can't tell you how many times I've connected to a caller only to find that they've forgotten to turn on their mike or turn up their speakers.)

   On the PC, you can check your microphone by running the Sound Recorder (sndrec32) program. You'll most likely find it under your Start menu in the Accessories/Multimedia section. With the Sound Recorder, you can not only record your voice to check whether your microphone and speakers are working, but also gauge how close or far away you should be from the microphone to produce clear audible sound. Hundreds of people I've talked to speak either so close to the microphone that they overload it, or are so far away their faint speech is drowned out by the whirring of their disk drive. By recording your voice, and adjusting the volume control on your microphone, you can hear just how loud you will sound on the other person's speakers and judge how not to blow out the other person's eardrums with your, "HELLO, HELLO, CAN YOU HEAR ME?"

   If you don't hear your voice after you've tried to record it, then either the microphone is not on, or it's not plugged in, or another piece of software is controlling the audio input.

for mac users *With a Macintosh, audio input can be switched easily between the microphone and CD player. Check your Control Panel/Sound/Sound In option, as shown here, to make sure the audio input is set to the microphone and not the CD player:*

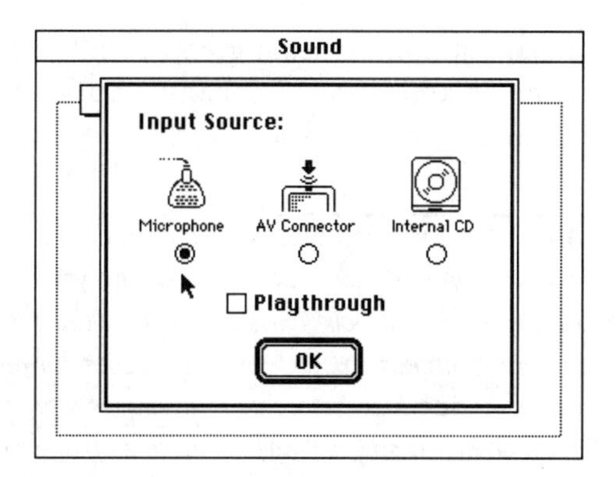

3. Make sure you have turned off the TV, radio, CD player, or whatever else may cause background noise.
4. Make sure your microphone is in a safe place. It should be a good distance from your speakers so it doesn't pick up what the other person is saying. You should also avoid positioning the microphone too close to your computer, or else it may pick up the grunting and groaning coming from the machine.
5. Think of something to say other than, "How's the weather?"

Funny, isn't it, that although you've used a telephone all your life you've never just picked it up hoping to randomly dial and start chatting to someone you've never met, and absolutely don't know.

That's what makes Internet phones so interesting. Sure, they save you money, but they also bring people together in a random and fascinating way. You'll find people from all over the world, from all backgrounds, and all socioeconomic levels. Some have been using Internet phone programs for a while, and others are real newbies, struggling to make their first connection. So be prepared to answer some mundane questions at first. The weather is especially a big topic of conversation. But be inventive. Bring along your list of favorite links to talk about, or study up on world news, so your Internet phone conversations can be more than just, "Hello, can you hear me?"

OK, Now You Can Really Use the Phone

Now that you've downloaded FreeTel, adjusted your microphone, connected to the Internet, and prepared a stack of entertaining things to talk about, you're ready to make your first call. FreeTel is a server-based product, so you don't have to know much more than how to double-click the program's icon. FreeTel handles the rest. When you first run the program, a screen will appear asking you to fill in some basic information, such as your name and what user name you would like to be identified as while connected to the FreeTel server.

remember *The first time you run FreeTel, be sure to click the Options button to specify what type of modem and sound card you are using. Also, almost all Internet phone programs will walk you through configuring the software to your machine's settings and Internet connection. Each program varies, however, and some may ask for your e-mail server name, so before you run any new program, be sure to have that information handy so you'll have a successful, uneventful installation.*

As soon as the program is configured, you're ready to connect. It may take a few minutes for the program to contact the server and then find and list everyone else who is using the program. Shortly, however, you should see a list of people's names scroll down the center of your screen.

People who are currently engaged in conversations are noted with an asterisk (*) before their name. Scroll down until you find someone whose name doesn't have an asterisk before it. Double-click the name of the person you want to talk to, and in a few seconds you should be connected to that person.

It's best to wait a few seconds before saying hello. It takes that long for your equipment to sync up with the other person's. Say hello and hear what happens. If the person on the other end is as ready as you are, you should hear a reply. If not, and if you've adjusted all your equipment according to my handy checklist, then there's most likely a problem with the other person's equipment. If after a few minutes you still don't hear anything, you can use the text chat board to type a message. The other person will see what you type and be able to enter a reply.

note *In some programs you have to hit the ENTER key to display your message on the other person's screen. You don't have to do this in FreeTel. Exactly what you type will be shown on the other person's screen immediately. Keep in mind that if you're using a real-time chat feature like FreeTel's, and you type something like "Turn on your mike, for gosh sakes! What is WRONG with you?", you won't have a chance to edit out your frustration before sending the message.*

If you've tried text chatting, and you've waited for the other person to do an equipment check but you still don't hear anything, let them know that you're hanging up, and then try calling someone else.

for mac users *VocalTec's Iphone connects in much the same fashion as FreeTel. Iphone is probably the best program to try first if you are a Macintosh user and want to start chatting quickly.*

You are now well on your way to chatting up a storm with thousands of people across the world, and becoming totally addicted to the Net. One regular chatter, Philip P. in Colorado, tells me that he now has new friends whom he talks to every week. He's replaced watching television with communicating, and it has opened up a whole new world for him.

And There's More to Internet Phones Than Just Shooting the Breeze!

Each Internet phone program has its own set of features that you can use to enhance your voice communications. Let's take a look at some of the most common available tools.

Text Chat

Text chatting lets you send typed messages to the other person, which are either received by the other user immediately as you type, as in the preceding FreeTel example, or sent in whole after you press ENTER or click the Send button. Each program's chat feature is a little different, but the concept is the same. You type. They read. They type back.

Text chat is a great feature to use if the other person can't hear you or doesn't have a microphone. It's also great when you need to communicate with the other person but the connection isn't that great.

Another way that people use text chat is to give out their Web page or e-mail addresses. Instead of saying, "OK, it's h-t-t-p colon slash slash" while the person you're talking to takes frantic dictation, you can simply type it out so the other user can see it. Some text chat areas, such as NetMeeting, let you cut and paste text, making it even easier to pass along favorite Web pages.

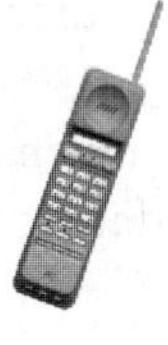

tip *Speaking of NetMeeting's text chat feature, it also lets you save your text conferences to a text file. This is a great feature for businesspeople who may need to keep track of what's been said online.*

File Transfers

You mean you can transfer files while you talk? Why would you do that? Well, let me give you an example. My dear Mom is a clip art fanatic. I recently purchased a CD with a great collection of color clip art. Mom needed a picture of a Cupid, which I just happened to have. We met on FreeTel one night, and I used the File Transfer feature to send the clip art to her.

But businesspeople also use this feature to quickly send spreadsheets, memos, advertising flyers, and all sorts of electronic files. Brad P. from New Jersey used the feature just recently to help another Iphone user upgrade his sound card drivers to full-duplex. "It worked great," reported Brad, "and within minutes the guy had a full-duplex system without having to go to the FTP site!"

Whiteboards, or Shared Paint Programs

A picture is worth a thousand words. This is especially true if you are trying to give directions or explain something, or if you just want to play a game of tic-tac-toe. Shared whiteboards are basically painting programs that let you and the other caller paint, draw, and write collaboratively.

This may sound like a rather silly waste of bandwidth, but think about how many times you would have liked to *show* the person on the end of the telephone line what you were talking about. Whiteboards make that possible. (Of course, I use mine regularly to play tic-tac-toe with my friend John M. in New Zealand.)

Application Sharing

NetMeeting was one of the first products to offer an application sharing feature. Although it requires a fast connection, working collaboratively over the Net is truly one of the more interesting and constructive ways to communicate with other people.

Say you want to show Microsoft's Help Desk people the error message you get when you run Excel. Simply share the application, and the person on the other end will see on their screen exactly what you see on yours, error message and all.

And you can pass the control over to the other person. This means that you and your accountant can interactively review figures and update spreadsheets, and see the immediate results on your screens.

As the bandwidth and speed of Internet connections increase, shared applications will make telecommuting a realistic and practical work option.

Wasn't That a Blast?

Making your first connection really isn't that complicated, is it? OK, so it's a bit more work than using your regular telephone, but when you compare the small amount of time and frustration with the money you can save, plus the added bonus of being able to meet interesting people from all over the globe, Internet phones just can't be beat.

TELEPHONE

Chapter 8

Gadgets, Gizmos, and Other Great Things for Internet Phones

YOU'VE got the sound card. You've got the microphone. You've got the software. What else do you need? Internet phone gadgets, of course. Whether it's a nifty noise-reducing headset or just a gizmo that keeps noise out of your telephone line, it's listed here.

Today you can choose from a ton of gadgets for Internet telephony—from microphones that look like telephone handsets, to sound cards for sound-challenged laptops, to speaker/microphone combinations that stick in your ear—and every day more and more of them appear on the market. You'll find them all described in this chapter, plus, in the back of this book you'll find special discount coupons for some of the products described here.

Telephones Masquerading as Microphones and Speakers

SoundXchange: A Phone? Maybe Not

Interactive Inc.
Phone: (605) 363-5117
E-mail: sales@iact.com
Home Page: http://www.iact.com
Models: A, AX, and VC
Platform: PC
Cost: $69–$189

It's a microphone. It's a speaker. It's a phone. It's SoundXchange. And it looks just like a real phone (see Figure 8-1). This speaker/microphone telephone look-alike can be mounted on the side of your monitor. A plug for connecting your external speakers is positioned on the left side of the

FIGURE 8-1 SoundXchange can mount right onto the side of your monitor.

phone base, which allows you to hear computer audio when the SoundXchange isn't in use.

I found the sound quality to be lower than that of the average speaker, but better than most headsets. The real kicker is that it makes talking on an Internet phone just like talking on a real phone. And it provides you with a certain amount of privacy by redirecting the sound from those blaring megawatt speakers of yours to the speaker in the handset. It also comes with an on/off/volume control so you can quickly adjust the volume as it's coming through the handset.

SoundXchange can also be used as a regular microphone to record sound files and to annotate word processing documents, or, with the appropriate software, to issue voice commands to control your computer as you now do with a mouse. It works with virtually every sound card on the market, and makes the Internet phone experience seem more like you are really talking on the phone.

Not Comp-U-Phone: There's No Doubt, It's Not a Phone

General Telephone Company
Fax: (905) 840-7919
E-mail: gentel@vrx.net
Home Page: http://www.vrx.net/compuphone.not
Models: 2000; 3000
Platform: PC
Cost: $59.95; $119.00

Those bright guys at General Telephone Company in Ontario figured out how to turn a phone into a speaker and microphone. The Not Comp-U-Phone 2000 is basically a microphone and speaker inside a real, modified telephone handset.

The Not Comp-U-Phone 3000, on the other hand, is an entire phone that you plug into the back of your sound card. When the Not Comp-U-Phone is on the hook, your computer operates normally, with the sound going through your speakers. But when you lift the handset the sound coming from your speakers is transferred to the handset.

TransPhone: Swipe and Talk

Firecrest
Phone: 0171-409 1214 (U.K.)
E-mail: sales@transphone.co.uk
Home Page: http://www.transphone.co.uk
Platform: PC
Cost: Under $100 U.S.

Sure you can use this clever little handset to hear and speak to others on the Internet (see Figure 8-2). But one feature separates this product from the rest. It connects to the serial port of your computer and includes the standard numeric keypad found on a regular telephone.

FIGURE 8-2 *The TransPhone comes with a credit card reader on the handset.*

On the back of the handset is a smart-card or credit card reader. You can use the phone not only to talk to other people, but also to pay for purchases over the Net with a swipe of your card. Look to see this type of phone handset being used more and more for those who would rather swipe than type their Visa or Mastercard number.

Head Gear and Headsets

The Quiet Zone 2000: Getting in the Zone

KOSS Corporation
Phone: (800) 872-5677
Home Page: http://www.koss.com
Platform: PC or Mac
Cost: $199.95

Koss offers some nifty fold-up headphones that provide superior sound quality, albeit at a price. The reason for the cost? A special "noises-off" activation button that can eradicate any external sound hovering around you.

The Quiet Zone 2000 folds up into a nice little package that stores in the accompanying carrying case. A little expensive, but if you want to keep your conversations quiet, you already have a microphone, and if you travel frequently, this is perfect.

Jabra Ear Phone: It's All in Your Head

Jabra Corporation
Phone: (619) 622-0764
E-mail: info@jabra.com
Home Page: http://www.jabra.com
Models: Various
Platform: PC or Mac
Cost: From $120

You just stick it in your ear and start talking. The Jabra Ear Phone (see Figure 8-3) picks up your voice through the little plug sticking out of your ear and transfers the sound from your sound card away from your speakers to the earpiece.

FIGURE 8-3

Jabra Ear Phone

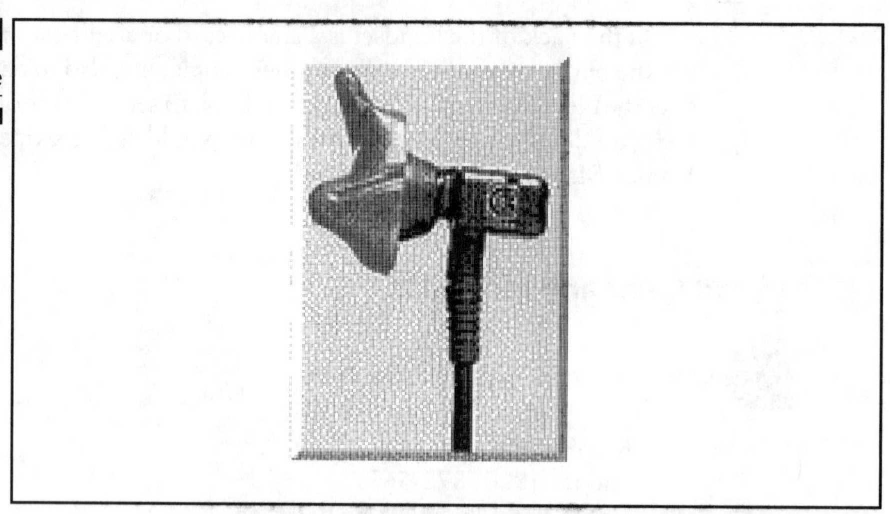

No headsets to give you a bad hair day, no microphone mouthpieces to get in the way of that morning doughnut. You stuff it in an ear and start talking.

It takes a while to get used to, with the earpiece being slightly larger than the in-ear headphones accompanying some personal stereos. The microphone is actually located in the part of the earphone that is positioned just outside your ear, but soon it just feels like a part of your head.

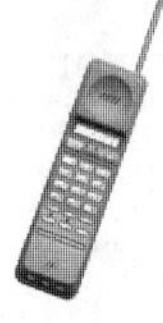

caution *With any headset/Internet telephone combination you should be very careful. Some Internet phones can occasionally emit loud bursts of sound. Make sure you keep the volume down until you are sure the person you are talking to has adjusted the gain on his or her microphone.*

Andrea Anti-Noise Headsets: Keep It Down Out There

Andrea Electronics
Phone: (718) 729-8500
Home Page: http://www.andreaelectronics.com
Models: ANC 100, 200, and 500
Platform: PC (Mac version is currently in the works)
Cost: $50–150

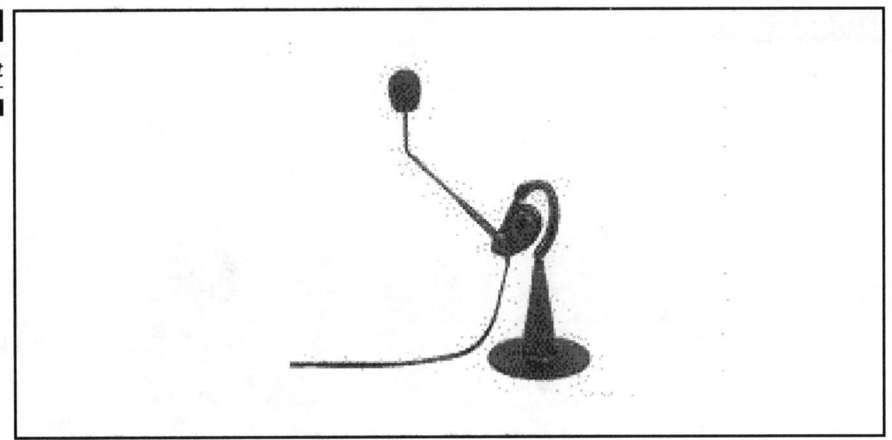

FIGURE 8-4

The Andrea headset

The patented Active Noise Cancellation technology incorporated into these earpiece/microphone headsets (see Figure 8-4) makes for crystal-clear voice transmissions. The earpiece fits snugly over your ear and the adjustable boom microphone fits a wide range of people.

Included is a switch that lets you use the headset as a free-standing microphone. Individual plugs for the headset and mike let you disconnect the headset speaker but still have hands-free conversations. An accompanying suction cup-like stand positions the microphone to any level you want. It's great if you want to continue working while you chat away with your Internet phone buddies.

Hello Direct's Monaural PC/Phone Headset: The $24.95 Solution

Hello Direct
Phone: (800) 444-3556
E-mail: xpressit@hihello.com
Home Page: http://www.hello-direct.com
Platform: PC or Mac
Cost: $24.95

A simple, low-cost solution for talking and listening to your Internet phone pals. The adjustable headset with a single padded earpiece fits snugly over your head (see Figure 8-5). The adjustable microphone rotates to a

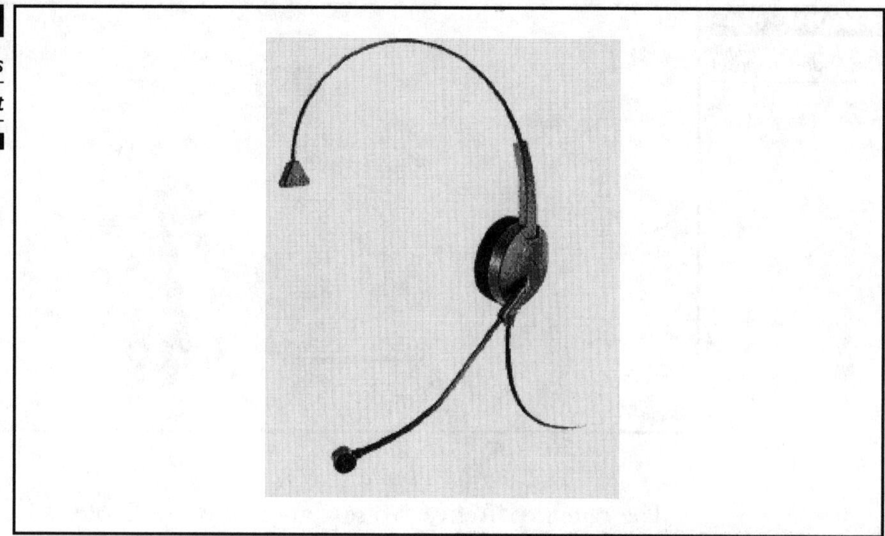

FIGURE 8-5

Hello Direct's monaural headset

variety of positions. You can plug both the headset and the microphone into your computer's sound card or use either separately.

Headset Accessories

CTI Adapter Cable: Adapted to You

Hello Direct
Phone: (800) 444-3556
E-mail: xpressit@hihello.com
Home Page: http://www.hello-direct.com
Platform: PC
Cost: $39.95

Have you grown attached to your Hello telephone headset? Were you bummed when you found you couldn't use it with your Internet telephone? Well, fret no more. The CTI Adapter Cable works with most sound cards, such as Creative Labs Sound Blaster, Media Vision Pro, or the MS Sound Card, to connect your Hello Direct telephone headset to your computer. One end looks like a telephone plug, the other end splits off into two jacks—one for the microphone, the other for the speaker.

The Headset Organizer: Simplify Your Life

Hello Direct
Phone: (800) 444-3556
E-mail: xpressit@hihello.com
Home Page: http://www.hello-direct.com
Platform: PC or Mac
Cost: $34.95

Do you keep losing your headset under that pile of papers on your desk? Then this is the perfect companion for any busy Internet telephone user. A clear plastic tray with headphone-hangerlike stand also doubles to hold your speakers, telephone, or paperwork.

Slide *The Internet Phone Connection* underneath and you've got everything you need to start talking on the Net.

Multimedia Keyboards

ConcertMaster Multimedia Keyboard: Sound at Your Fingertips

NBM Technologies
Phone: (800) 854-2465
Cost: $129

If you don't like bulky speakers hanging off the side of your computer, you should check out ConcertMaster, a full-size Windows 95-specific keyboard. It comes with a fully loaded 4-watt output speaker system featuring a frequency range of 50KHz to 20KHz, volume controls, a handy Mute button, along with a knob that lets you toggle between the keyboard speakers and your normal speakers.

An omnidirectional microphone is also part of the keyboard. The only problem with this type of speaker/microphone setup? If you speak softly you may have to up the volume on the microphone in the various Internet telephone programs.

Maxi Sound: Maximum Sound at Keyboard Level

Maxi Sound
Home Page: http://www.maxiswitch.com/www/maxi
Cost: $89

FIGURE 8-6

The MaxiSound keyboard with built-in speakers

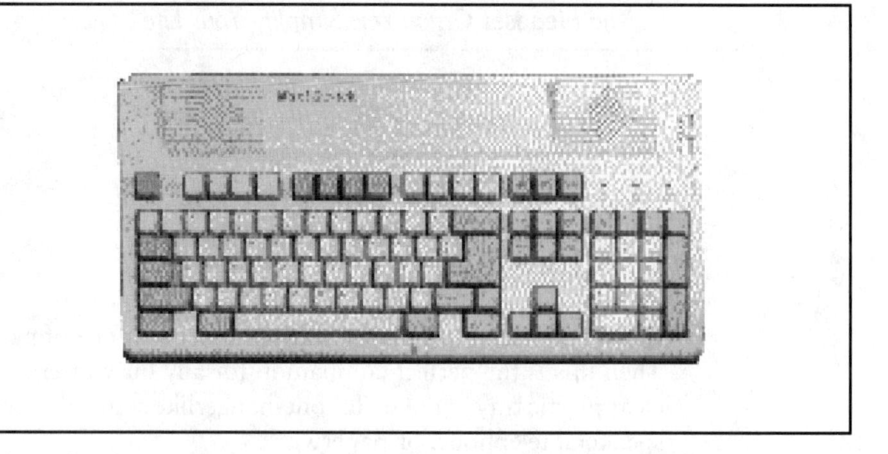

Equipped with Altec Lansing speakers, this keyboard provides excellent sound for Internet telephony applications (see Figure 8-6). A sliding volume control on the right-hand side of the keyboard makes for fast sound reduction when you need it.

The Maxi Sound contains a monaural microphone with an accompanying stereo input microphone jack. The position of the microphone, below the spacebar, works best for those who work close to their system. Integrated keyboards like this can certainly tidy up anyone's already cluttered desk.

Cameras

Connectix QuickCams: I See You, Can You See Me?

Connectix
Phone: (800) 950-5880
E-mail: info@connectix.com
Home Page: http://www.connectix.com
Platform: PC or Mac
Cost: $99–$230

Amazing. This little round ball offers 640×480 resolution in either black-and-white or color (see Figure 8-7). It doesn't matter whether you have a Mac or a PC; installation of a QuickCam is as simple as plugging in a modem.

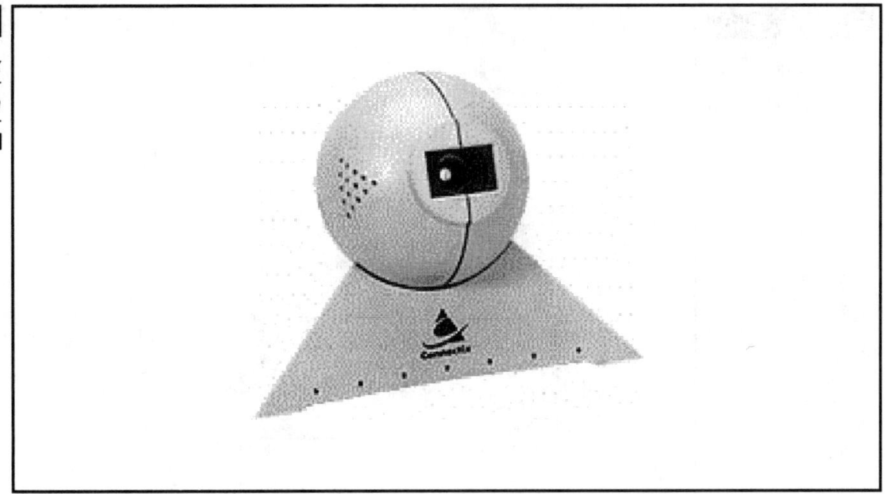

FIGURE 8-7 *Connectix QuickCam*

The color version can capture millions of colors, and the black-and-white comes with a built-in microphone. The camera works well with programs such as VDOPhone, FreeVue, and CU-SeeMe, and offers a relatively high-quality, built-in microphone. A hard rubber stand makes positioning the camera extremely easy.

You may want to hold off on the color version though. The video quality just isn't good enough yet, and the additional overhead to transmit color can slow down voice communications.

FlexCam: The Camera for the Flexible User

Video Labs, Inc.
Phone: (800) 467-7157
E-mail: videolabs@flexcam.com
Home Page: http://www.flexcam.com
Platform: PC or Mac
Cost: $395

It almost looks like some alien being, but FlexCam is basically a color camera on an 18-inch stick (see Figure 8-8). The third-of-an-inch camera offers a resolution of 500 by 582 pixels. The unit comes with two microphones and works with software for either PC or Macintosh computers.

Unlike the QuickCam that plugs into the serial port of either a Mac or PC, FlexCam requires a video capture card (unless you have an AV-equipped Mac).

FIGURE 8-8

Video Tabs' FlexCam

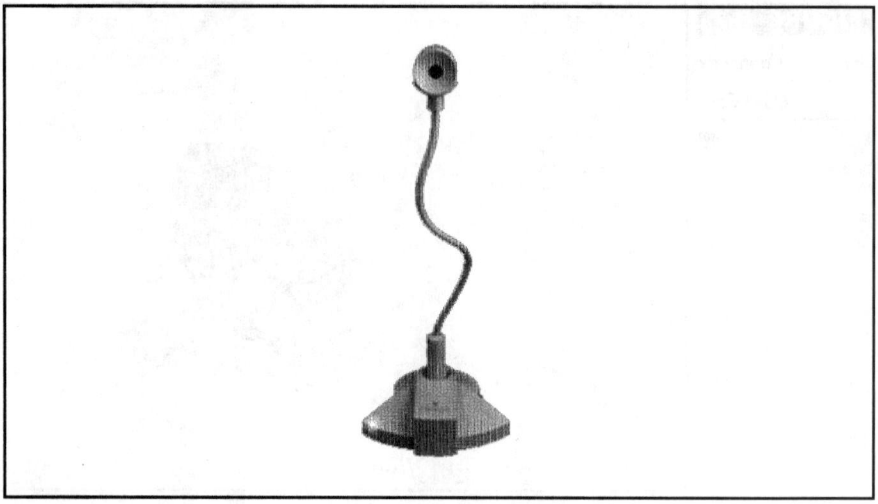

VideumCam: An Out-of-This-World Camera

WinNov
Phone: (408) 733-9500
E-mail: info@winnov.com
Home Page: http://www.winnov.com
Cost: $249–$500

With its colored base and extended lens, this camera closely resembles a tiny Martian (see Figure 8-9). It provides extremely clear color pictures when used in conjunction with an existing video capture card or the accompanying Videum Multimedia expansion card.

FIGURE 8-9

VideumCam kinda looks like a small space alien

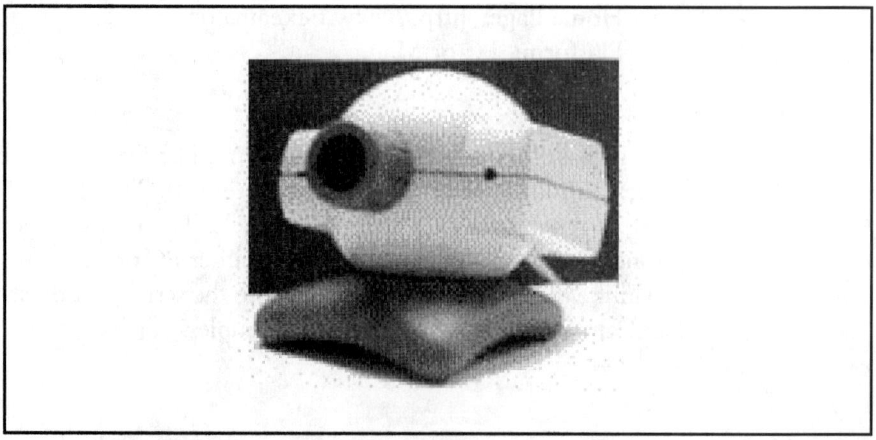

VideumCam is for use with VDOPhone and CU-SeeMe, providing crisp color pictures to your caller. Depending upon the video capture board used it can provide either full- or half-duplex sound.

Sound-Challenged Laptop Solutions

The CAT: Purrrfect for Adding Audio to Your Laptop

VocalTec
Phone: (201) 768-8893
E-mail: info@vocaltec.com
Home Page: http://www.vocaltec.com/portable.htm
Platform: PC
Cost: $199

Is your laptop lacking sound? Do you need something you can travel with that works with any laptop or sound-challenged PC? The CAT is the answer if you want a decent half-duplex sound card in a neat little box. The CAT plugs into any PC parallel port and comes with a speaker and microphone jack. Windows-compatible software drivers come with this product, offering full sound capabilities to not only your Internet phone programs, but to any other sound-enabled program as well.

Dad started out with this unit on his Compaq laptop, and he still uses it to keep in touch with Mom when he travels.

Media Vision Sound Card: A Clear Vision for PCMCIA Laptop Users

Media Vision
Home Page: http://www.mediavis.com
Platform: PC
Cost: $295

If you are looking for true 16-bit sound that offers full-duplex voice, a built-in microphone, and Sound Blaster capabilities—all in a PCMCIA plug-in card about the size of a fat credit card—this is a perfect product. It works with all laptops that have standard PCMCIA slots, and comes with all software to make it work with Windows 3.1 and Windows 95. This is just one of many PCMCIA sound cards now flooding the market.

Nex*Phone: The Next Best Thing to a Real Sound Card

Nex*Phone
Phone: (408) 727-6584
Home Page: http://www.inext.com
Platform: PC
Cost: $229–$436

Nex*Phone offers full-duplex sound in a tiny box that plugs into the parallel port of your laptop or desktop computer. It's great for laptops, IBM PS/2 MicroChannel-style computers, or regular desktop computers that may have a half-duplex sound card/modem combo board installed.

The various models offer everything from microphone, headset jacks, and battery-powered units to a package deal that includes Internet Phone from VocalTec.

Telephone Line Gadgets

Connect Protect: Keeping the Conversation Going

Public Address Marketing
Phone: (813) 824-8527
Home Page: http://www.advantag.com/777/connprot.htm
Platform: PC or Mac
Cost: $22

Is someone in your house constantly picking up the phone, forgetting that you are online? Then you need the Connect Protect series of telephone adapters. You plug the jack adapter into the wall where the telephone line is, then connect your modem line and telephone into this Y adapter. When your PC dials into the Internet, no other phones connected to the Connect Protect adapter will be allowed to interrupt your call.

RFI Eliminator: Do You Hear That?

Hello Direct
Phone: (800) 444-3556
E-mail: xpressit@hihello.com
Home Page: http://www.hello-direct.com
Platform: PC or Mac
Cost: $24.95

Do you hear a lot of static on your phone line? Is your modem constantly losing its connection because of a "dirty" line? Then you need an RFI Eliminator. You can get them for single- or two-line phones, and they will block interference from AM/FM and CB radios.

Chapter 9

Tips and Tricks for Optimizing Your Internet Phone Connection

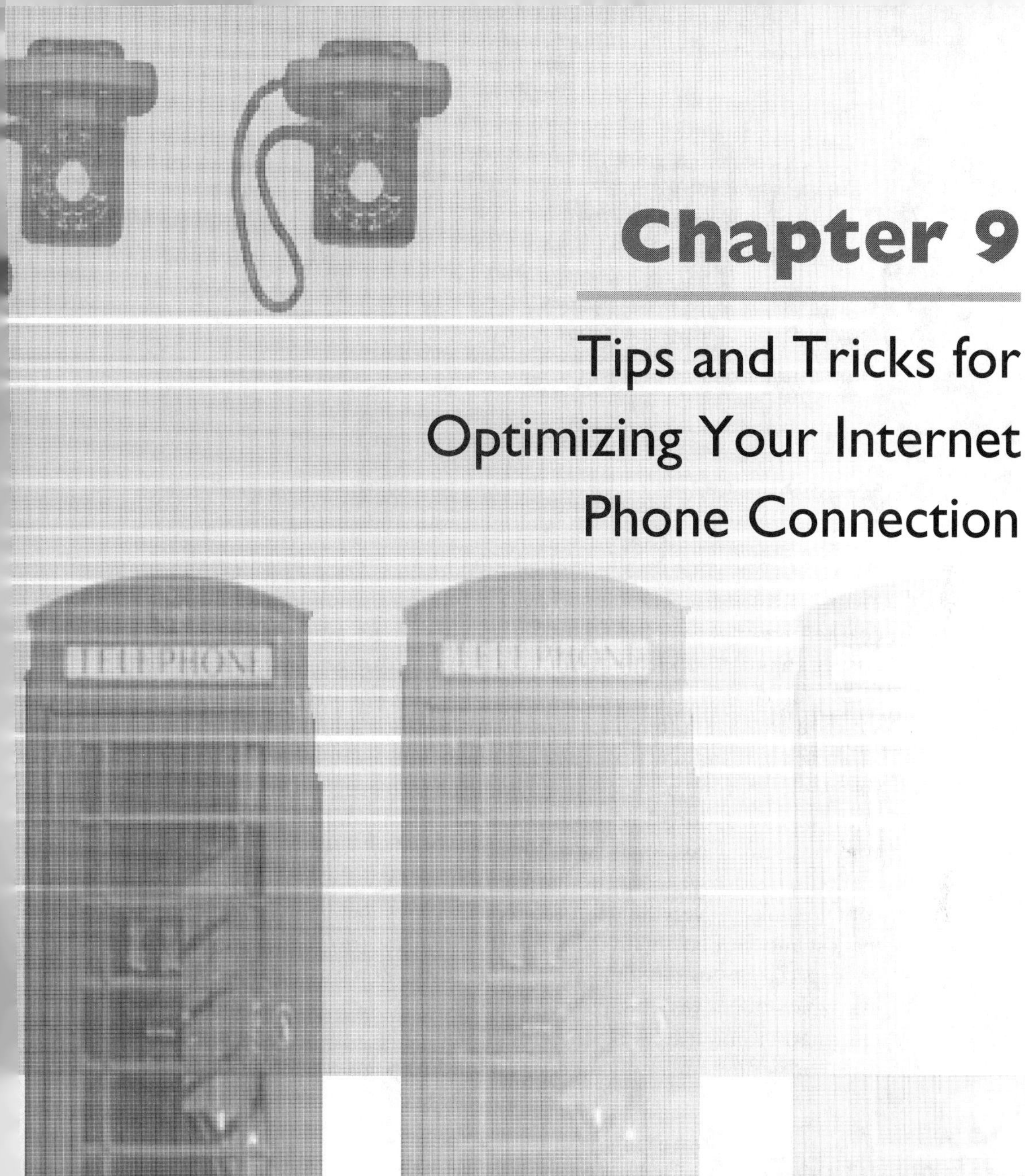

THERE are lots of ways to tweak your Internet phone connection. This chapter presents some tips that might help you get a better connection and crisper sound. I've included some places you can get shareware enhancements for your Internet phone programs, and I've also tried to point out some obvious, yet often overlooked, tips for the more popular programs.

These are just a few tips that might help you. I strongly encourage you to visit your program manufacturer's home page to keep up-to-date on changes in the program, download additions to technical support notes, and join in company-run discussion lists that link you to other users of the same product.

You can also visit my home page, **http://www.netphones.com,** which contains links to everything listed here, plus a few other items that may not have made it into the book in time. Feel free to e-mail me (**netphones@aol.com**) with your special tips or questions if you have them. You can also join my mailing list, which offers weekly tips, announcements of new products, and general information about making Internet phone programs really work. Instructions for signing up are included on my home page.

Connection Tips

Check to See If You Really Are Running an Internet-Connected Computer

The first key to making sure you have a good Internet phone connection is making sure that you really are working on an Internet-connected computer. Some providers, such as America Online (AOL), may not connect your computer directly to the Internet, nor assign your computer its own IP address. Instead, what providers like AOL do is connect your computer to their computer, turning your computer into a dumb terminal that simply issues commands to the AOL computer. The AOL computer does all the processing and sends you back the results. In essence, this means that the sound coming from an Internet phone program goes through the AOL computer, and not directly to the computer you're trying to communicate with.

If you want to find out whether your connection to the Net is a true Internet connection or just another "dumb" terminal, incapable of running IP-required programs like most Internet phone programs, try running WINIPCFG from Windows 95:

1. Connect to the Internet with your connection software.
2. Go to the Start menu and choose Run.
3. In the Run box, type **WINIPCFG**.
4. Hit ENTER.

If you see an IP address of 0.0.0.0, it means that your provider does not assign IP addresses. It also means you aren't really connected to the Internet as an independent computer that can take and transmit Internet phone calls from IP-dependent programs such as PGPfone.

Connecting Directly to the Net? Want to Know Your IP Address?

Again, you can run WINIPCFG to find this out:

1. Connect to the Internet with your connection software.
2. Go to the Start menu and choose Run.
3. In the Run box, type **WINIPCFG**, and then hit ENTER.
   You should see your IP address listed in the IP Address section of the dialog box.

remember *If an Internet phone program asks you for your e-mail address, that's not the same as your IP address. An e-mail address is simply a pointer to a location where your e-mail will be sent. It's not actually the location of your computer. That's why many Internet phone programs, such as PGPfone and the Direct-Dial feature of Intel Phone, require an IP address—they provide point-to-point connections and don't go through name servers to convert the e-mail address to an IP address.*

When in Doubt, Get a Fixed IP Address

If you can coax your Internet service provider into giving you a fixed IP address at little or no additional charge, go for it. Having a fixed IP address is like having a fixed phone number. Unlike a dynamic IP address that changes each time you log on to the Net, much like having to jump from phone booth to phone booth, a static IP address lets people call you directly. The caller doesn't have to worry that your address changed.

Post Your IP Address to a Web Page So Your Friends Will Know Where to Call You

Try the nifty, swifty little $20 shareware program called Online. It lets you easily post your dynamic PPP Internet address to an HTML document, allowing others to look up your current Internet address and contact you. Information, including the time you logged on to the Internet, is posted in a formatted HTML document.

You can download it at **http://www.4dcomm.com/~rrhubott/programs/online/oline151.zip** or check the official Online home page at **http://www.4dcomm.com/ ~rrhubott/rhusoft.html**.

Check Your Service Provider Settings

Programs such as DigiPhone require that you know the *exact* name of your mail server. Oftentimes service providers only give you an alias for the mail server, with the actual name of the mail server computer being something entirely different.

I found this out when I absolutely couldn't get DigiPhone to work, no matter what I did. Finally a call to DigiPhone's tech support clarified that my e-mail server, which is **mailhost.alaska.net**, is actually served up on a machine called **byatt.alaska.net**.

Anyway, the tech support guru figured this out by using a program called Trumpet Dig, a shareware program available to anyone via Trumpet's home page at **http://www.trumpet.com**. It's not a particularly fancy program, nor does it come with a whole host of instructions. But if you're interested in fixing your own problems, it might offer some insight into how various Internet service providers have their systems configured.

You basically give Trumpet Dig a domain name and it gives you back a list of devices currently being served up through that host. Interpreting what Trumpet Dig is saying, however, takes more knowledge of networks than most people have. But by comparing several domains just about anyone can start to get a feel for what Dig is trying to tell you.

Why Is Our Conversation So Delayed?

Do you know how many routers, computers, and network servers your voice has to travel through in order to get to the intended caller? If you did, you'd probably be amazed that you even get a good connection.

TRACERT is a program that can show you exactly how many network "hops" it takes to get from your server to another server. It will also show you if any of the network connections along the way are having problems, and indicate how long it will take to get to your intended destination.

To run TRACERT from Windows 95:

1. Go to the Start menu and choose Programs.
2. Select the MS-DOS prompt. A black window will come up with a C:\ prompt flashing at you, waiting for you to type in a command.
3. Type in **tracert** and the domain name of the server you are trying to contact. For example, if I wanted to know how many hops it takes to get to my Dad's server, I would type **tracert flex.net.** (Flex.net is my Dad's server.) You can also enter the IP address.

   In a few seconds a list of computers, routers, servers, and gateways will be listed, showing you how many hops it takes to get from here to there.

It's Not Flashy, but It Can Do the Trick Too

If you don't have Windows 95 and still want to run TRACERT try **http://www.ict.org/~nelgin/utilities.html**.

This page will let you run TRACERT and other Unix-related utilities to check your connection speed plus the distance traveled between your service provider and another domain. All this is done via a standard Web page form.

You can also Ping (or send a 1-byte packet to) the computer you are trying to connect to, just to see if the computer is up and running. (Ping was discussed in detail in Chapter 4, just in case you are wondering, "What the heck is Ping?")

remember *When you check connection response time from other computers or domains, there could be problems or delays on the computer you are using as a starting point. Although response times are a good indicator of network traffic, other variables can affect overall speed.*

Getting More Details About the Quality of Your Connection

If you want to see what kind of packet loss you might be experiencing or what kinds of delays are slowing down your conversations, browse on over to the interactive Web page from Onlive! Technologies: **http://www.onlive.com/cgi-bin/nettest.cgi**. This page tests your connection every ten seconds and records its findings over five-, ten-, and twenty- minute averages.

Upgrade to a 28.8 or 33.3 Modem

By far, the best enhancement you can get is a faster connection to the Net. If your budget limits you to dial-up access as opposed to direct connections through ISDN or cable modems, make sure you get the fastest modem possible, and make sure your service provider can match that speed.

The 28.8 and 33.3 Kbps modems are the fastest dial-up connections available today. But remember, you have to check with your service provider to make sure your modem speed matches theirs.

Changing 28.8 Modem Speeds to Boost Speed

Although Windows 95 sets up most 28.8 modems just fine, there are occasions when it doesn't and sets the maximum speed rate lower than what the modem is expecting. In the case of all 28.8 modems your maximum speed should be set at 57,600. Here's how to make sure of this:

1. Click the Start button and choose Settings, Control Panel.
2. Double-click the Modem icon.
3. Click Properties (below the white box).
4. For a 28.8 Kbps modem, choose 57,600 from the "Maximum speed" box. For a 14.4 Kbps modem, choose 19,200.

Having Problems with Your New 28.8 Modem and Windows 3.1? It Could Be the Driver.

The Comm.drv communications driver supplied by Windows 3.1 doesn't support speeds above 19.2. So if you have a 28.8 Kbps modem, and you've experienced freeze-ups, random disconnects, or very choppy voice quality, update your communications driver, as follows:

1. Get the latest shareware updated driver at **http://oak.oakland.edu/simtel.net/win3/drivers.html**

2. Once you download the driver, you should copy it to your Windows\System directory.
3. Now you will need to edit your System.ini file. In the [boot] section, look for Comm.Drv= and replace it with **Comm.drv=Cybercom.drv.**
4. Make sure you reboot your machine after making these changes.

Try Modifying the TCP/IP Packet Size in Windows 95 to Improve the Performance of Your Internet Connection

The quality of an Internet connection can be affected by many things, such as traffic (both local and between you and the other caller), the type of connection, and the other programs running over your Internet connection.

Changing the TCP/IP packet size in your Windows settings might help, but it will mean delving into RegEdit to modify a few things, so if you don't feel comfortable tweaking your system this tip isn't for you.

remember *You should always have a backup of your Registration Database before you start making changes. And always remember to restart your computer after you make any changes to your system files.*

To change the packet size, follow these steps:

1. Start up RegEdit by going to the Start menu and typing **regedit,** then hitting ENTER.
2. Once you are in RegEdit, locate the following directory and click on this directory name: HKEY_LOCAL_MACHINE\System\CurrentControlSet\Services\VxD\MSTCP\
3. From the Edit menu, create this key (or modify it if it's already there):

   DefaultRcvWindow

4. Set or type in a string value around 8192.

Getting an "Unable to Load Winsock.dll" Error?

If this error is bogging you down, do the following:

1. Make sure your Internet connection is up and running.
2. Make sure Winsock.dll is included in the Path statement in your Autoexec.bat file.
3. If you have multiple Winsock.dll files, make sure the correct one appears as the first entry in your Path statement.

note *Although you can have multiple Winsock.dll files in your system to connect to different providers you should only work with one version of the Winsock.dll file. Do not use multiple Winsock.dll files from different Internet providers, or from online services such as CompuServe, America Online, or Prodigy, unless you are using one of these providers to log on to the Internet.*

Making Sure You Have the Latest Trumpet Winsock

If you are using Trumpet Winsock, make sure you use the latest version (currently 2.1). Some products don't work with older versions of Trumpet Winsock and Tcpman.exe. Although you may be getting your e-mail and browsing the Web just fine, Internet phone programs are much more picky about the type of programs you use to connect to the Net.

You can get the latest version of Trumpet Winsock and other Winsock utilities from **http://www.trumpet.com.**

remember *Make sure you are using the Internet dial-up connection provided with Windows 95 or the Windows 95 Plus disks. Other versions of the Winsock.dll may not work properly with Windows 95.*

Automatically Connecting to Your Service Provider Without Going Through the Dial-up Network Dialog Box

Try QConnect for Windows 95. This program will dial your connection, automatically log on to the Internet for you and then close itself. You can download it at **http://netnow.micron.net/~jjordan/dbs/qconnect10.zip** or check out Qconnect's home page at **http://netnow.micron.net/~jjordan/dbs/qconnect.html.**

Using Commercial Services

NetCom Is Not Winsock-Compliant

NetCruiser 2.0 and earlier versions are not Winsock 1.1-compliant. This incompatibility prevents some Internet phone programs from operating properly. Make sure you get the latest NetCruiser connection software from **http://www.netcom.com**.

Enabling Winsock in NetCruiser

You need to enable the NetCruiser Winsock support before you can use Internet phone programs with NetCruiser. When in NetCruiser, you enable Winsock support by doing the following:

1. Start NetCruiser and establish a connection (session).
2. Go into the Settings menu and choose Startup Options.
3. Click "Autoload NetCom's Winsock.dll."
4. Click OK.
5. Exit and restart NetCruiser; Winsock will now be enabled.

America Online (AOL)

AOL offers support for direct Internet connections; however, you need to download a special Winsock.dll file:

1. In AOL, go to the Keyword field and enter **winsock**.
2. Follow the instructions provided, or check **http://www.aol.com** for the latest information on connection drivers.

You Can Always Use Trumpet Winsock with AOL and CompuServe

Use the latest version of AOL's Internet dialer to connect to the Internet. But if that doesn't work, or if you get numerous connection errors, you can also use the Dial-up Networking option in Windows 95. In the case of Windows 3.1 or Windows for Workgroups 3.*x*, try Trumpet Winsock if your Internet phone program does not operate properly with AOL's latest Internet dialer.

Internet Phone Program Add-Ons

Speak Their Language, Pronto!

Pronto is a Windows program designed to help you greet your fellow Internet Phone users in their native language. You can download your copy from **http://www.halcyon.com/pronto/pronto32.exe**, or try the official home page at **http://www.halcyon.com/pronto/welcome.htm.**

IBM Connection Phone Doesn't Offer a Text Chat Feature. Is There Some Way to Add One?

You can always bring up WinTalk, a simple freeware program that lets two people text chat in real time. WinTalk supports the Unix Talk protocol, so you can talk to Unix users as well as anyone else running the Unix Talk software.

note *This is one of many text chat programs. For more TALK clients, check out* ***http://www.shareware.com****.*

You can download it at **ftp://ftp.elf.com/pub/wintalk/wtalk126.zip** or check out the official home page at **http://www.elf.com/elf/wintalk/wintalk.html#introduction.**

OK, Fine. But What About a Whiteboard? I Need to Draw Diagrams While I Talk to My Friend Bob in London

Try Wanvas' $20 shareware whiteboard program that lets several users draw with a variety of tools on the same picture from different computers.

You can download it at **http://www.owt.com/ember/zips/wvs11_32.zip**, or check out the official home page at **http://www.owt.com/ember/software.html.**

But I Want More!

Then you should try Internet Conference. The software not only lets multiple people view and mark up documents and images remotely over the Internet but lets them actively collaborate in developing and editing them. You can simultaneously view images and documents and annotate them. Find Internet Conference at **http://www.vocaltec.com**.

Specific Internet Phone Tips

Virtually every Internet phone provider offers technical support via its home page. So I would check there for additional information, tips, frequently asked questions (FAQs), and special trouble-shooting information. As these programs are upgraded, many of the problems originally encountered will go away. But for now here are some of the more common problems you might run into with specific applications.

Reading the White Pages in Intel's Internet Phone

If you can't launch the White Pages from Intel's Internet Phone with Internet Explorer 3.0., then change the page location (select Options, then Page Locations) for your White Pages by removing the final front slash from the path.

Old way: **http://www.IAF.net/**

New way: **http://www.IAF.net**

Read the Readme Files

Check the Readme files, FAQs, and the Newsgroups regularly for tips and hints on most Internet phone programs. Resolutions to specific customer problems can be found in these locations.

Getting Them to Shut Up in FreeTel

If the other person's microphone will never stop transmitting, you can hold down the CTRL key to override the other person's transmission.

Operating Behind a Firewall with FreeTel

FreeTel does not support Proxy servers so it may be difficult to get FreeTel to work within a corporate environment using a firewall. According to FreeTel's technical support, however, you can have your system administrator change the firewall to permit both incoming and outgoing UDP traffic on port range 21300 through 21303.

For security reasons, your system administrator or MIS manager may not permit this change. If opening that port is not allowed, one of the following may be permitted instead:

- Enabling the above change only for the IP address of your computer
- Connecting your computer directly to the "hot" Internet connection outside of the firewall

NetMeeting and Firewalls

The current port information used by NetMeeting is T.120 port is 1503. The ULS server listens for calls at port 12345.

Ask your systems administrator if he or she can allow these packets to pass through the firewall.

Want to Videoconference with NetMeeting? Learn to Share!

If you have a QuickCam, you can share the QuickPict software with the other caller. QuickPict displays what the camera sees in real time. By sharing the application the other party will see what you see as it's displayed on your QuickCam camera.

Hardware Tips

What's That Buzzing Sound?

If you hear your speakers buzzing, or if other people complain of a buzzing sound, check your speakers to make sure they aren't picking up electromagnetic interference. To reduce or eliminate buzzing noises, reduce the sound card's volume level and keep the speaker cables away from the power cords, especially the PC's cord. Also make sure you keep your speakers away from a poorly shielded monitor. Monitors can also interfere with speakers.

Why Is There No Sound Coming from Your Microphone?

The problem could be in the microphone itself, or within the wiring, or you could just have a corrupted sound driver. Here are some troubleshooting steps:

1. If your microphone has an On/Off switch, make sure it's turned on.
2. Check to see that you have the microphone connected to the correct plug at the back of the sound card. Remember, it plugs into the microphone plug, not the speaker plug!
3. If your microphone connects to an external speaker system that requires power, make sure the speaker power is turned on.
4. Check to see if the microphone's volume control needs adjusting. Do the same for the speaker volume.
5. Check the mixer program in your computer. Make sure you have the software controls turned on and that the volume level is not too low.
6. Once you're sure that your microphone is in good working order, find out the name of your sound driver through your mixer program, and reinstall that driver.

Pump Up the Volume

If your sound system just doesn't move your woofer or excite your tweeter, then try NuReality's Vivid 3D Plus add-on for your sound card and speaker set. It's a small box that attaches to your sound card and reprocesses sound signals. This recompressing turns just about any old tin-canny sound card into an awesome high-quality system. The box plugs into your sound card and speakers and can be installed by anyone with a screwdriver. It's available for PCs and Macs for the low, low price of only $99.95. NuReality can be reached at (800) 501-8086 or (714) 442-1080.

Position Your Microphone Correctly

I can't tell you how many times I've connected to someone and instead of hearing a clear, crisp "Hello," have heard something more like someone trying to talk while they stand next to a 747 ready to take off. If there is one bit of advice I can give you, it's to position your microphone close enough to you, but far away enough from the computer, so that the other person doesn't hear your disk drive spinning or the fan whirring all the time.

Sensitive microphones and those set with their volume settings on high can pick up all sorts of noise, including what's sitting next to them, like the computer.

But Not That Far Away!

Don't you hate it when people call you on speakerphones? It's like they're too lazy to pick up the telephone handset, or your conversation just isn't as important as the busy work they're doing on the other end.

Well, if you don't position your microphone close enough to your mouth, the other person won't hear you very well and will feel like they are trapped in speakerphone hell. Plus, they'll also hear everything that is going on around you. This constant drone of noise can overrun the default sensitivity settings on most Internet phone programs, causing your program to think you are talking all the time, and thus not letting the other person get a word in edgewise. To make sure you have a good connection, don't invite friends over to play the piano, dance, and sing all night while you try to talk on an Internet phone.

If Feedback Increases on Your Side, Their Side, or Both

This usually means your microphone is too close to the speakers, or the other party's microphone is too close to their speakers. The microphone detects the sound coming from the speakers and sends it back to the other party, causing unwanted feedback.

To work around this, you can increase the voice activation level and decrease the speaker volume on your computer.

If you are using a full-duplex sound card and the Internet phone program has a voice detection setting, have them adjust it until the feedback goes away.

If that doesn't work, either move the microphone away from the speakers or use a headset.

What if the Other Person Can't Hear You?

Did you forget to turn on your microphone? Did you forget to plug it in?

Is the microphone and/or voice activation level set too high or too low? (You can usually check it on some programs by a "Test the Microphone" feature.)

If the other person's voice activation level is too low or too high, you'll be put in a constant Listen mode. Tell the other person to adjust the level up or down. Some programs allow you to override the other person's setting. Check the tech support page of the manufacturer for specific information on how to do this.

Long Delays in Speech

The Internet may be experiencing some delays, thus causing delays in the conversation. You might be able to change the compression setting to allow for slowdowns within the program you are using.

You or the other party may be experiencing packet filtering problems. The best solution is to disconnect from the Net and try reconnecting.

If Your Computer Slows Down After a Few Minutes

Make sure you aren't running any other programs, such as printing or updating Web pages.

Also, some programs allow you to record conversations, which will slow down your machine considerably. Check to make sure you aren't recording.

Lastly, periodically clean out your temporary folder or directory, and defragment your hard disk regularly.

If the Sound Is Very Scratchy

Make sure you aren't running a DOS session or formatting or copying disks. Anything that accesses the hard drive, such as having Virtual Memory turned on with the Macintosh, can also cause breaks in the sound. Check your Control Panel settings to make sure Virtual Memory is turned off, and that you aren't using RAM Doubler, which can also slow the conversation down.

Heavy network traffic can also delay conversation and make it sound garbled or scratchy. Your only option is to hang up and wait until the network traffic slows down or call during off-peak hours—in the early morning, midday, or very late at night.

You might have set the sound compression too high, forcing the computer to work faster to compress the sound. Adjust it lower and see if that has any effect on the sound quality.

If you are using the SoundBlaster16 or AWE32 with Creative Labs' full-duplex driver, try replacing it with the standard, half-duplex driver.

If the Program Hangs Up During Your Conversation or Stops Working Entirely

Hey, get a clue. Maybe the other person just decided to hang up on you. Think of more interesting topics to discuss besides the weather and try again.

But seriously, the computer could have received an error message, so it may be waiting for you to prompt it to continue. This happens more in

Windows 3.11 and on the Macintosh than in Windows 95. Check to see if there are any error messages waiting for your okay, then continue.

You could just be experiencing heavy network traffic. Use Ping or TRACERT to see if you can reach the other computer.

Your computer may simply not be powerful enough to handle the program. Check the minimum qualifications and compare those to your computer.

Your system has crashed or is on the verge of becoming unstable. Some programs will actually check to make sure all system resources and hardware are working properly. If they aren't, they will exit out of the program immediately to avoid a major system failure. If you get a General Protection Fault in Windows, or a Serious System Error on the Macintosh, try restarting the computer, rebuilding the desktop on the Mac, or deleting the temporary files and running ScanDisk on the PC.

If the Program Operates Only in Half-Duplex Mode

Do you have a half-duplex card? Do you have the updated drivers for the card? If you don't know who manufactured your card go to the Settings/Control Panels/System device in Windows 95 and check the Device Manager for the properties of your sound card. Or check the hardware specifications of your machine in your user's manual.

No Sound Coming from Your Speakers or Headset?

Is the volume turned up? And is the sound output specified for your speakers? Check the Sound Control Panel on the Mac or the Sound Settings on your PC.

Is another program running that is currently controlling your sound card? If so, exit the Internet phone program and the other program that is controlling the sound, and restart your Internet phone program.

Are your speakers plugged into your sound card? Check the connections to make sure there is no problem with the speaker connection.

What if You Can Place Phone Calls but You Can't Receive Any?

Do you know how to answer the phone on the particular Internet phone program you are using? Do you have to click a button to answer? If so, don't delay; click that button today.

You may also be experiencing packet filtering problems. Basically, this is when certain packets may not be let through your firewall or through the service provider's system. You will need to contact your Internet service provider or network administrator to determine if packets addressed to the following ports are passed through to your side.

FreeTel Settings

TCP 18794 "inbound" and "outbound"
UDP 18794 "inbound" and "outbound"
TCP 18793 "outbound"

WebPhone Settings

TCP 21845 "inbound" and "outbound" WebPhone-to-WebPhone control packets
UDP 21845 "inbound" and "outbound" WebPhone-to-WebPhone audio datagrams
TCP 21846 "outbound" connections to NetSpeak Directory Assistance Server
TCP 21847 "outbound" connections to NetSpeak Connection Server

You could also experience the same thing in reverse, in which others can call you but you can't call them. The settings for packet addressing are the same.

Where Do You Get Creative Labs' Full-Duplex Drivers?

You can find them at **http://www.creaf.com**. Look for the latest upgrades available in the tech support section.

Getting Rid of That Echo on the Other End

If you can hear yourself talking during an Internet phone conversation, as if there is an echo, it probably means that the other party is using a full-duplex audio card and you are using a half-duplex card.

In order to stop the echoing, the full-duplex user should try the following:

- Increase the voice activation level
- Decrease the speaker volume
- Keep the microphone as far away from the speakers as possible
- Manually turn off the microphone when you are finished speaking
- Try using a headset with a built-in microphone

If the Talk Settings Flip from Talk to Idle Even When Someone Isn't Speaking

Depending on the Internet phone program you are using, it may be that either the program is picking up feedback from your microphone, thus "keying" the mike as if you were talking, or your voice activation level is not set properly.

This could also happen because you are using a half-duplex sound card and the other party has a full-duplex card, and either his or her microphone is too close to the speakers, or the voice activation level has been set too low for the microphone.

If Your Computer Freezes

Turn off the full-duplex operation in your Internet phone program or from the control settings of the sound card. Sometimes the Creative Labs full-duplex sound driver can cause problems with certain Internet phone programs. Also, check to make sure that you have installed all the sound drivers in the proper directories, if additional drivers were included with your Internet phone program.

Can You Connect a Different Microphone to Your Older-Style Macintosh?

You can if you get Apple's Line Input Adapter Cable. One should have been included with your computer, but if not, you can call (800) SOS-APPLE, or check their Web page (**http://www.apple.com**) for product support.

Do You Spit When You Say Sassafras?

You can get windscreens or pop filters for your microphone. Check the local Radio Shack or your local music store. Windscreens prevent air currents coming from your mouth from abruptly striking the microphone, which can cause popping and thumping noises.

Want Better Voice Control?

Buy a microphone with an on/off switch. That way you can quickly turn on and off the mike without having to make software adjustments.

Wait a Minute. I Know I Turned the Volume Down on My Microphone Yesterday!

Some programs actually reset the volume and sound controls you may have set. If you've run any multimedia games, or voice recognition or sound recording programs, make sure to double-check your voice settings before starting your next Internet phone session.

General Etiquette

- Make sure you have all the parts working before you make the call.
- Don't be rude to people or hang up on them. Remember, you never know—you might meet on the street tomorrow someone whom you talked to on the Internet phone last night.
- Don't yell into the microphone. Speak at a comfortable level. If the other party can't hear you, then either adjust your microphone or VOX level, or talk louder.
- Don't play music, turn on the TV full blast, or have people chitchatting over your shoulder while you talk. The person on the other end can oftentimes hear every background noise, sometimes even better than your voice.
- Watch out for people who want to engage you in cybersex chats or videoconferences. If you happen to connect to another person who wants to have cybersex and you don't, be polite, and then let them know you are going to hang up.
- Don't fire up the latest video chat program to see "what's happening" while your children are in the room. You might find that they get an amazingly quick sex education lesson on the first call. Make sure you know who you are connecting to before you drag little Johnny and Janey into the room. Your mother would be shocked and surprised to find what people are doing with videoconferencing these days. So was I.

- If you do feel the urge to engage in cybersex, one bit of caution: Find out the age of the other participant first. Enticing a minor into sex acts, even in cyberspace, is a crime—and really disgusting, low-life behavior too.
- Don't transmit files to users without their permission, and don't accept a file unless you know what it is and what it is supposed to do. If you do get a file, run a virus-checking program against it to make sure the other person isn't spreading viruses. Make sure you get the e-mail address of the person who sent you the file so you can inform them if the file indeed does have a virus.
- Don't pull up whiteboards or other shared programs without first telling the caller what you plan to do. Having a game of hearts pop up on someone's screen can be confusing, especially if they have never seen that capability before or don't know how to play hearts.
- If you can't hear the other user, and a text chat feature is available, bring it up and let them know you can't hear them.
- Wait for your turn to talk. Most people think these things operate just like phones. They don't. They operate more like speakerphones or CB radios. Give the person a chance to say what he or she wants to say, then speak. Some people use "over" when they are through speaking, as you would with a CB radio. I prefer the wait-and-talk method myself. Pacing yourself can really make a conversation much more cohesive.
- Think about more to talk about than the weather. If you are interested in Spain, come prepared with specific questions for your long-distance caller.
- Fill out any greeting information appropriately. If you want people to call you because you want to talk about goat herding, then include that in your greeting.
- Pick the appropriate chat room for the type of conversation you want to hold. Don't go into the "Speaks Korean" chat room if you can't speak a lick of Korean.

General Internet Phone Tips

Check Out Newsgroups and Mailing Lists to Keep Up-to-Date

Many of the telephony software providers have a mailing list or Web discussion group available. Check the Web page for the manufacturer of the program you use; you may find special instructions on how to sign up for one of these mailing lists.

You can also check out the various newsgroups available. Here is just a partial list of some of the more useful newsgroups pertaining to Internet phones:

- alt.Winsock.voice
- comp.os.ms-windows.apps.com
- comp.os.ms-windows.apps.Winsock.misc
- comp.speech

General File Downloading Tips

If you have Windows 95, you need the latest version of WinZip (6.2) for Windows 95. You can get it at **http://www.winzip.com.**

The latest version is fantastic. It will do the following automatically:

- Search your drive for zipped files
- Run installation and setup programs
- Let you specify which folder you want the decompressed files to end up in
- Delete temporary files from your drive, if you choose

Program Cannot Load—System Says It Can't Run Two Copies of the Same Program

Programs such as Trumpet Winsock need to load first before any Internet programs start up. Many programs, such as VocalTec's Internet Phone and

IRIS phone, are specified to autoload each time you start Windows. This can cause a problem, since the Winsock DLL may not have fully loaded by the time these programs startup. This will cause the Windows error "Cannot start more than one copy of the specified program" when you try to start these Internet phone programs again and reconnect to their servers.

Here's the best course of action in this situation:

1. Remove these programs from your startup group or Startup folder.
2. Replace them with Trumpet Winsock or the Internet dialer of your choice.
3. Set up the new dialer so that it loads first.
4. When you want to use your Internet phone program, load it manually.

Tips for Getting Better Sound

Adjust Automatic Gain Control

Make sure you have Automatic Gain Control (AGC) turned on. AGC, a feature inherent in your sound card, boosts the microphone sound level automatically when you talk. It also reduces the level of background noise it picks up when you aren't speaking. Each sound card's AGC is turned up in a different way. Consult your sound card manual or ask the manufacturer where to find this feature. It may also go by the name of 1 SRS-3D Control. Oftentimes you'll find it in the mixer control program.

Make Sure the Compression Matches Your Machine

You can adjust the codecs or compression schemes used in some programs, such as Speak Freely and PGPfone. Check the Sound Compression chapter for the codec that most closely matches the type of Internet connection and computer you have.

I can't tell you how many times someone has said to me, "That program just sucks. I couldn't hear a thing." The problem wasn't in the program, but rather in the compression settings. Many programs that let you manually set compression schemes don't automatically pick the best compression scheme for your machine the first time you run the program. Instead, you have to match the compression scheme to the machine and speed of the Internet connection you are using. Take some time to tweak this feature and you'll find that those programs that didn't sound so hot now offer excellent quality and response times.

Want Good Sound? Don't Run Other Applications While Using an Internet Phone Program

Almost all Internet phone programs let you run other applications and other Internet services such as retrieving e-mail, file transferring, and Web browsing at the same time the Internet phone program is running.

However, when you run other programs you are taking away some processing power that the Internet phone program may need. The same holds true for utilizing other Internet services. When your Internet connection's bandwidth is fully used, or when your computer's processor is running full-speed, you may experience halting or even loss in the conversation. Usually this drop in quality is temporary and decent sound quality will resume when the system is back to running only the Internet phone program.

Miscellaneous Tips

Getting More Screen Space

Want to get the status screen off the bottom of your Netscape window so you can see more of Intel Phone and the Web page? Hold down the CTRL key and press S. This will toggle off the status bar, giving you more screen "real estate" so you can see more of the Web page while Intel's Internet Phone program is running.

Want to Know What Country a Caller Is Calling from?

Try the shareware program called *Country Code Converter*. This simple program takes a domain and spits out the country's name. For example, enter in SE and the program will answer "Sweden."

You'll find it for downloading at **http://www.intermotion.com/ctfire/ccc10.zip**.

What If You Just Want to Send a Quick Note Telling Someone Your IP Address?

Definitely try NetPopup. NetPopup is a Windows 95 utility for sending short notes over the Internet in real time. There's no e-mail server delay and no need to bring up a separate application just to send a single-line note. The other person has to have NetPopup running as well. It's shareware, but reasonably priced at $10 per copy.

You can download it at **http://www.vtoy.fi/~malo/proges/netpopu2.zip** or check the official NetPopup home page at **http://sik.ppoy.fi/~malo/netpopup.html.**

You can also try NetNote, a simple little freeware program that generates and sends "Post-it"-type notes and alarms to other people across the Net.

You can download it at **ftp://www.process.com/ftp/pub/win95/nn32_18b.zip** or check the official NetNote home page at **http://ourworld.compuserve.com:80/homepages/vetty/.**

Chatting Away All the Time and No One Else Gets to Cruise the Web?

Vicom lets you share one modem between two Macintoshes, creating a gateway of sorts to let two people use the Internet at the same time. It does take some effort, but you can have two Macs running in about an hour, accessing the Internet through a single modem and single user connection. Check out Vicom Technology's home page at **http://www.vicomtech.com.** Costs run about $150 per two-user license.

Is Windows 95 an Internet Barrier?

Tired of being stopped dead in your tracks when you start to log on to the Internet while using Windows 95? Get the Don't Stop utility, which bypasses the connection dialog screen and gets you connected directly to the Internet. You can find it at **http://www.shareware.com.**

Stay on the Net with Ponger

Want to keep your Internet connection up and running with Windows 95? Try Ponger. You can find Ponger at **http://www.shareware.com.** Ponger will send a data packet to your provider on a regular basis, thereby maintaining a continuous connection to the Net. This is useful to have when you have a dynamic IP address and have given someone the number, but you need to wait until they call you. Ponger will maintain your connection so you don't have to keep e-mailing them a new IP address.

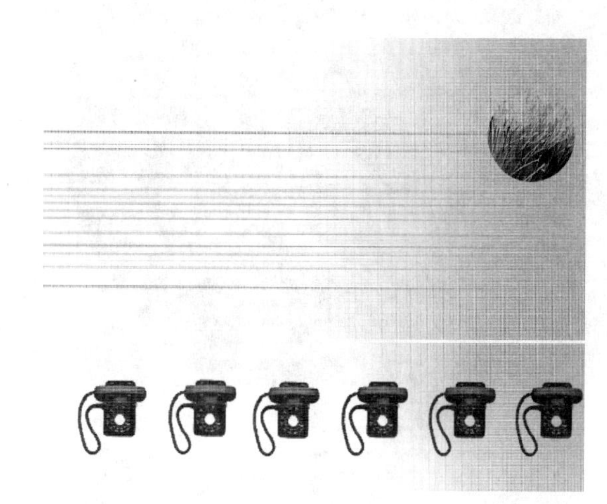

Chapter 10

More Telephony and Audio Products...

CHAPTER 6 didn't contain everything. Yes, I know it's hard to believe, but there's more! The products listed in this chapter deviate slightly from the true definition of Internet telephones. Each product listed here either offers additional features that makes it *more* than an Internet telephone, or offers a different form of "live" video or audio. Some of them are videoconferencing programs, and some connect your computer to the telephone networks, while others are really collaborative application-sharing products. Still others don't provide interactivity between two users, but do offer live voice and/or video, much like a TV or radio.

All of these applications represent tremendous programming creativity. Their authors truly understand that the Internet is not just about a bunch of "dead" Web pages; instead they seem to realize people want to see each other, collaborate with one another, and at times, just be "Net potatoes" staring at a screen, watching a little TV, or listening to their favorite music on the radio. The companies who make these products should be applauded for their ingenuity.

For me, these products were by far the most fun to test and review because they did so much. But be forewarned. Multifunctionality, the ability to deliver audio and video at the same time, means fast computers and fast connections. You should pay careful attention to the minimum requirements listed, and realize these are *minimum* requirements, meaning your Macintosh IIci may be able to display video with CU-SeeMe but may choke when it tries to transmit audio. Your 486 might work fine text chatting with PowWow, but falls short when you try to talk to someone using the audio portion of the program.

Also, these programs, more than any other type of Internet programs, work best when the Internet is not busy. To see them really shine, I recommend you try connecting with them early in the morning or very late at night. If you can't do that, then take into account that congested network traffic can cause many of these programs to have choppy sound or slow refresh rates, just like Internet phones.

Categories

I've classified these programs into five different areas of functionality. Within each category is a listing of some of the more popular and functional programs and, also, a brief description of each. This listing doesn't include every single interactive program in the universe. Rather, it's a glimpse of what is available for communicating live with other people. I listed the best I found, but there are lots of others, and more coming out each day. If you'd like to see what else is out there, drop on by the "Official Netphones" home page at **http://www.netphones.com.**

The five categories are as follows:

- Video phones
- Multiparty conferencing
- Text chat programs for those computers that are multimedia challenged
- Internet-to-phone gateway applications
- One-way multibroadcast streaming audio and video

The following sections explain each of these product categories, and the actual list of products comes after that. I've only reviewed in detail products that provide two-way audio interaction. For the rest, I've provided a brief rundown of what they do and what nifty features they include. The little CD symbol next to a product means its programs are on the CD, so there's no need to waste time downloading them. I've included the addresses to the home pages for products that aren't on the CD.

note *On my Web page (**http://www.netphones.com**) you can subscribe to the "Official Netphones" mailing list, which lists announcements of new Internet telephone and interactive Internet products, plus tips on how to make the various Internet phone programs work for you, as well as some ideas on how people are putting these programs to good use. To subscribe via e-mail, send an e-mail message that includes your name and e-mail address to **netphones@aol.com**.*

Video Phones

Remember in the 60s and 70s how those futuristic TV programs always talked about how the world would someday be able to communicate with "picture phones" or "video phones," forever leaving the simple telephone behind? Well, videoconferencing seems to have caught on, but not through the telephone networks, but rather on the Internet. As a matter of fact thousands of people are videoconferencing on their desktop computers right now, with relatively good sound and video quality. They are using computers equipped with cameras and microphones to talk to people all over the world, just like those TV shows predicted.

Video phones are software programs that combine audio *and* video in one package. They work with a video camera attached to a video capture card and a sound card to offer incredible quality in video and audio (especially when you consider the network traffic and the relatively slow computer systems most people deal with each day). I've included more products in this category than any other because, although they are different, they are close cousins to Internet telephones.

If you've ever seen them demonstrated or used them yourself, you know that using a video phone is *way* cool. You feel like Captain Kirk on the bridge of the Starship *Enterprise* talking to other life forms across the galaxy. (Of course having the last name Kirk helps, too.)

One downside to using video phones is that too few people have cameras, and/or fast enough computers to display really quick-moving pictures. It may take a while before you find someone to hook up with who has both a camera and fast computer—*or* someone who is willing to talk to you if you don't have this equipment! Although most people with cameras are patient and will let you check out how their picture looks, many are just as anxious to see other people as you are, and they may cut you off quickly if you don't have a camera.

remember *Parents, beware! With video phones you can really SEE the person you're talking to, and more importantly what the other person is doing, and who or what he or she is doing it with or to. You'd be surprised how many people feel compelled to take off their clothes and talk dirty to total strangers. I've seen some things that would make most sailors blush, so be careful. Above all, don't gather the kids around the computer screen oblivious to the fact that people use video phones for video sex. When you see something like "Wanted! Female in a bra to call me for a quick flash!!" you should assume they probably aren't talking about flash bulbs. As a matter of fact, one software manufacturer, FreeVue, realizes the potential of children "dropping in" on adult conversations, and requires you to register through a Web page confirming that you are indeed at least 18 years of age.*

One thing you'll notice quickly is that the quality of the picture and sound varies with the type of camera and capture board people use. Today many people are using the Connectix QuickCam camera because it's so cheap—it runs about $100. You might think of investing in one of these little round marvels if you're short on cash. Even if you don't have a camera, do check out video phones. You'll be amazed by the technology, and sometimes astounded by what you see.

Multiparty Conferencing Programs

Conferencing programs are mainly a combination of text chatting, application sharing, and audio options. However, some programs, such as PowWow and NetMeeting, have more options than just the standard conferencing features. Many conferencing programs allow more than two people to type or join in on the conversation, though the audio portion in most programs is usually restricted to allowing only two people talking at the same time, with the exception of 3-D OnLive!.

Conferencing is great because it opens up a whole world of personalized chat rooms. Unlike using chat rooms in America Online, you don't have to pay connect charges outside of your standard Internet connection fees, *and* you can create your own chats, both public and private, that can be as outlandish as you wish. The chat world is all yours—there are no censors.

Conferencing chat programs are amazing when you consider that most of the people who take advantage of conferencing applications don't really know each other. You're certain to meet a lot of different people with varying views from all over the world. Somehow, although they are spread across the globe, chat programs create a certain feeling of community, albeit virtual.

One word of caution, though: Some of these conferencing programs can be a bit confusing at first, so I highly recommend you read the documentation or help files of the program you intend to use *before* you use it. Some folks have a tendency to just jump right in and forego reading any how-to instructions, but if you don't do the reading, I can guarantee that you'll miss out on some pretty nifty features.

Chat Programs (for Audio and Visually Challenged Computers)

I've also included a couple examples of interactive chatting programs, programs that let you type real-time to another person connected to the Net. These are great programs for those of you who own computers that may be audio and video challenged, but still want to experience "live" conversations on the Internet. Tons of these programs exist. Most are variations of the Unix Talk command, and/or may connect to the Internet Relay Chat channels, or IRC for short. As a matter of fact, one of them may have even been included in the software you got from your Internet service provider when you first signed up. If you have a Mac, you might have been given a program called Homer; PC users might have WinChat. Each one of these programs can connect you to another person and let you type text back and forth to each other.

Once you get the "live" bug, if you don't already have a sound card and speakers, I'm sure you'll want to add multimedia capabilities so you can "hear" the people to whom you've been typing messages.

Internet-to-Telephone Gateway Connections

Sounds nutty doesn't it? Trying to connect a Web page to a phone, or a phone to an Internet user. But connecting the Internet to the telephone system is really a great idea. Just think. You could update audio messages you've placed on your Web page from the airport in London. You could use your computer and Internet connection and an Internet phone carrier to call a business associate who isn't on the Net, and save a bundle. And just consider how easy it would be to pick up the phone and send a voicemail, fax, and e-mail message all in the same call.

Today you can do all that, and more. Companies such as IDT and GXC offer extremely discounted rates on long-distance calls if you use your multimedia computer, Internet connection, and their software to connect to their telephone networks. You can dial any telephone number in the world. The person on the other end picks up their telephone and you start talking, microphone to telephone.

This category is hard to describe, because you have to see it to believe it and understand what each application has to offer. The benefits usually boil down to saving money.

If all of this seems very esoteric to you, I recommend you start with the easiest of the bunch: IDT's Net2Phone to make long-distance calls with your computer. Once you try something like this, you'll start to see exactly how exciting this marriage of the two technologies can be.

One-Way Stream Audio or Video Products

The products in this category aren't true Internet telephone, but "live audio." In essence, however, that's just what Internet telephones are: live audio.

These programs bring radio and TV to your desktop. What's so great about this technology is that you can be in Racine, Wisconsin, listening to the local radio station in Hong Kong. All you need is the Real Audio player and the address of the radio station you want to listen to; then you're rocking out.

Or you might want to drop in to see the latest movie trailers from Finland with StreamWorks. And if you're not a night owl and you never get to see the late-night news program, *CBS Up-to-the Minute*, you can always catch it on VDOLive any time, day or night.

OK, so not every bit of content is live, but these programs, like Internet telephones, do have one thing in common: They open up your eyes and ears to the goings-on and people scattered all across the globe. And when you consider that virtually all of these streaming audio players are free, you can't beat that.

Video Phones

CU-SeeMe

White Pine Software
Home Page: http://www.wpine.com
Platform: Mac, PC
Cost: Commercial version under $70 from White Pine; beta version free from Cornell University.
Rating: ☎☎☎☎

Minimum Requirements

- 486DX/66MHz PC
- 8MB RAM
- 10MB free disk space
- 256-color (8-bit) video with 640×480 or higher resolution
- Windows 95, Windows 3.1, Windows NT, Windows for Workgroups 3.11
- Video camera if you want to send video
- Sound card and microphone if you want to hear audio

or

- Mac with a 68030 processor and 25MHz clock

- MacOS System 7
- 8MB RAM
- 10MB free disk space
- QuickTime version 2.0
- Sound Manager version 3.0
- Speakers or headset if you want to receive audio
- MacTCP version 2.0.6 or Open Transport version 1.1
- Video camera if you want to send video

Features

CU-SeeMe (see Figure 10-1) was originally developed at Cornell University for their desktop videoconferencing needs, and it's still available free of charge from Cornell's site. The enhanced version is sold as a commercial product through White Pine Software. CU-SeeMe is desktop videoconferencing software for real-time person-to-person or group conferencing. Both Macintosh and Windows users can connect and see each other, exchanging live pictures, sound, and text over the Internet in real time. People can connect to each other via IP addresses or can conference with multiple users

FIGURE 10-1

CU-SeeMe in action

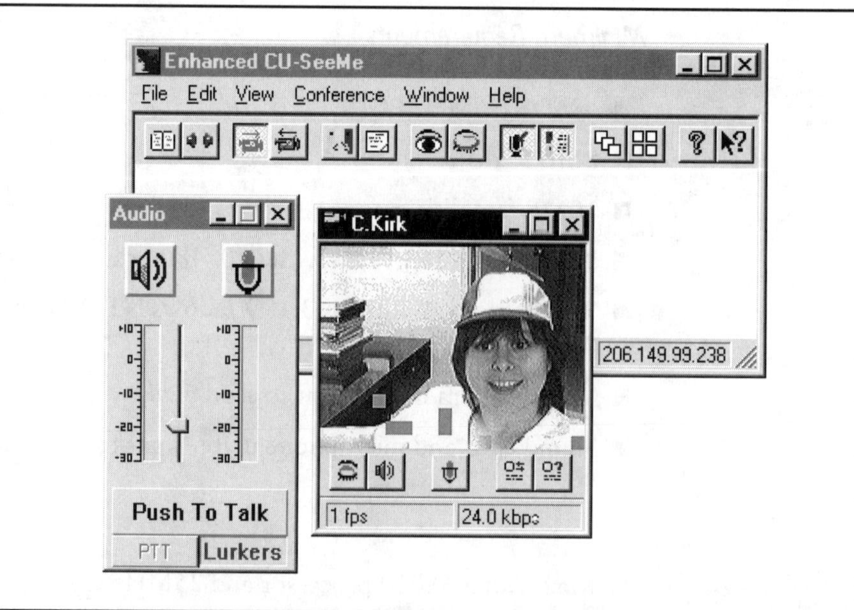

via "reflectors" or computers that run the CU-SeeMe server-type software that allows multiple people to connect at a single source.

The Cornell version offers black-and-white video, with decent audio, and the ability to take snapshots or "slides" that can be saved and presented as a slide show to other viewers. The White Pine version offers full-color video (provided the camera used is color) along with an enhanced audio feature, a collaborative whiteboard, plus the standard text chat feature found in the Cornell version. Both versions of CU-SeeMe can be launched directly from a Web page with your favorite Web browser, and you can view up to eight people at a time, and talk via text and audio to as many people as you wish.

The software supports both 14.4 and 28.8 Kbps modem connections, with about a 1 to 3 frame-per-second refresh rate for a 28.8. However, you can adjust the settings to get better compression based on your machine and Internet connection capabilities. You can add the names and addresses of those you frequently connect to through the use of the built-in "phone book" feature. In addition, you can add reflector sites where people go to conference to the reflector list. If you connect to private conference reflectors you can password-protect these entries.

Personal Opinion

This is what got everyone started, including myself and my family. It's a great product, and there is always lots of activity on the reflectors, or CU-SeeMe channels. NASA regularly broadcasts shuttle missions and, when not in orbit, educational programs. I recommend CU-SeeMe to anyone, even if they don't have a sound card and camera, since the program still lets you tune in to the various reflectors.

One problem, though, is if you have a slow connection and a slow machine. You'll most likely be able to see the video, but the audio will come in broken and—usually—nearly inaudible. If you've been using Cornell's free version of CU-SeeMe, the commercial version provides better audio quality and color; however, if you use the full-color feature, expect slow refresh rates. That means that you may see really jerky movements, resembling those pictures of the first space missions to the moon more than a modern TV broadcast. If you ever have a chance to see the program in action on a fast link, you'll be amazed at how clear the audio is, and how quickly the pictures refresh. Even on the local Internet café's 56K link, the difference was definitely noticeable.

Outside of special broadcasters like NASA, most of what you'll see on the public reflectors are people staring back at you, and late at night, people exposing themselves. On NASA's site, however, I've seen amazing pictures of astronauts walking in space. And this year during the early presidential primaries, I stumbled across Pat Buchanan giving a campaign victory speech in Massachusetts.

During the late evening hours, however, expect some raunchy text chat, so keep the kids away, unless you know there isn't any deviant behavior going on. And check both Cornell's and White Pine's home pages for event listings. Many nationwide community events, like the Republican National Convention, are being broadcast over the Internet via CU-SeeMe. Although it isn't TV quality, you do get to see things you may never see on TV.

I've only had about two or three decent conversations—not because of the people, but because of the poor audio quality and slow response when typing text messages. Unlike Internet phones, videoconferencing with CU-SeeMe with a slow dial-up modem seems more suited for experimentation and fun than for real communication. However, if you have a fast link—something around a 56K or above—or are on a company-wide network, CU-SeeMe would be the perfect tool for desktop videoconferencing.

FreeVue

FreeVue Telecommunications Network
Home Page: http://www.freevue.com
Platform: Windows (Macintosh version coming soon)
Cost: Free
Rating: ☎☎☎✆

Minimum Requirements

- 486/33MHz PC
- 4MB RAM
- 3MB free disk space
- Sound card if you want to hear audio
- Microphone if you want to talk
- Video capture card and camera or QuickCam camera if you want to transmit video

note *The current version of FreeVue does not work with CompuServe, although the manufacturer is working on this problem. People who work behind firewalls need their administrators to punch a hole (UDP port #2222) to allow FreeVue to operate.*

Features

FreeVue software lets you send and receive live video and audio over the Internet with amazing clarity and refresh rates. FreeVue is much like CU-SeeMe, but with a few different features (see Figure 10-2). With FreeVue, multiple people can view, talk, and text chat with each other. The client software connects to the FreeVue server, then sends your special nickname, or "handle," to a Web page that lists other currently connected FreeVue users. At present, unlike CU-SeeMe, there is no direct IP to IP address calling, but if you know the handle of the person you want to call, all you have to do is type that in, and press the Call button.

You can also create private password protected logins to your FreeVue-enabled computer to keep unwanted callers from calling you. This means only the person with the secret password will be able to connect to you automatically. A single Bye button lets you disconnect an individual caller from a conference, and you have control over the number of calls you can simultaneously take. You can use the text feature to chat with other users.

Using a standard 28.8 Kbps modem, you can expect a refresh rate of 2 to 5 video frames per second, and up to 10 frames over an ISDN line. If you don't have a camera you can still use the audio portion of the program to talk to other FreeVue users. The video refresh rate for

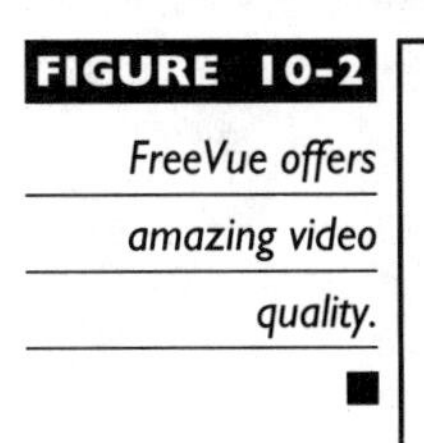

FIGURE 10-2

FreeVue offers amazing video quality.

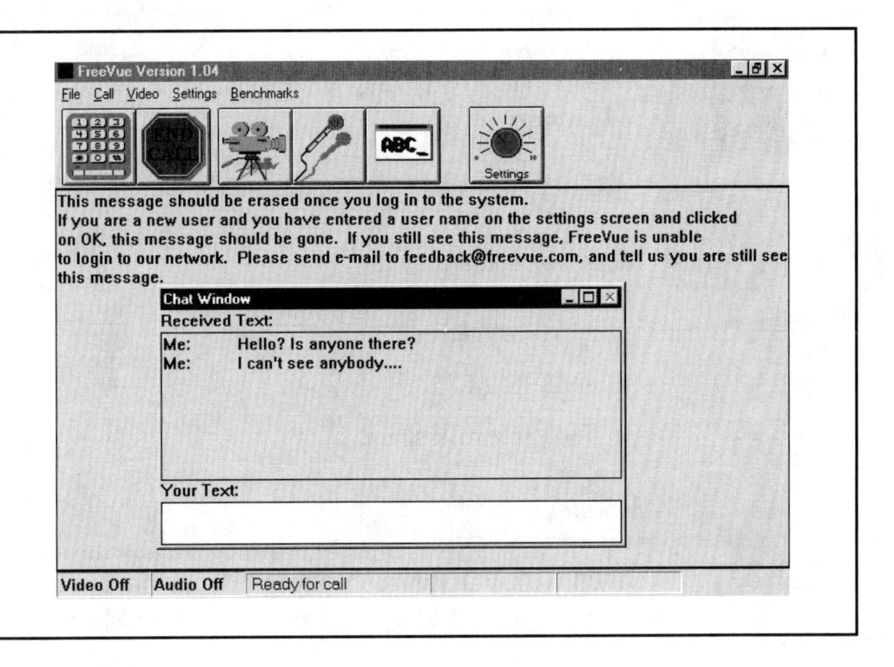

dial-up modem users is far better than CU-SeeMe, but the picture quality is a little less clear.

Personal Opinion

Wow! I personally think the refresh rate is amazing, giving very smooth video. However, the voice quality is somewhat disappointing at times. I could certainly hear what people were saying, but oftentimes the sound was very muffled, like they were holding a handkerchief over the microphone. I wouldn't recommend using FreeVue entirely for voice communications, although you could get by if you had a relatively clean connection to the Internet.

I wouldn't say that everyone who uses FreeVue is into cybersex, but the vast majority of the names you see listed are looking for people who want to bare it all in front of the camera or be titillated by what other conference callers say or type. The fact that you have to be 18 years old before you can see that list of conference callers tells you this product is used by an adults-only audience.

The interface is a little simplistic, and the fact that you have to jump from Web browser to FreeVue to call other users was a little awkward, but overall the program is relatively easy to use. Just don't use it in front of the kids.

VDOPhone

VDOLive
Home Page: http://www.vdolive.com
Platform: PC
Cost: Free
Rating: ☎☎☎☎✆

Minimum Requirements

- 60MHz Pentium
- 8MB RAM
- 5MB free disk space

Features

VDOLive's VDOPhone lets you talk to and see any other VDOPhone user (see Figure 10-3) as long as you have a camera, a sound card, and a

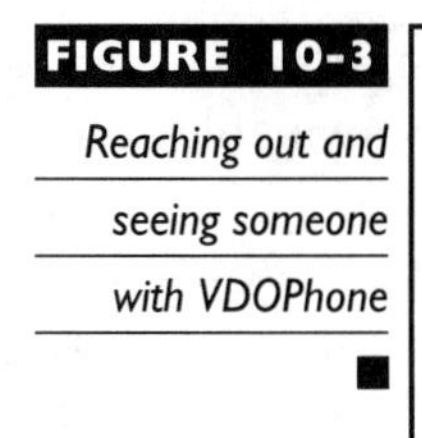
FIGURE 10-3 *Reaching out and seeing someone with VDOPhone*

14.4 Kbps (or faster) modem. You'll need a relatively fast computer. Don't expect to run VDOPhone with much success on a 486 or less.

VDOPhone offers not only video, but audio and text chat features. If the person on the other end does not have video capabilities, you can include a small .gif or .bmp picture that will be transmitted to the other caller when you first connect, along with a scrolling text message. You can also magnify the picture to twice its normal size.

To find other VDOPhone callers, you use your Web browser to view a Web page-based listing. This listing can be pulled up directly from within VDOPhone, or you can e-mail an invitation to chat via VDOPhone to a prospective caller. Adjustable audio and video controls let you increase the volume on the microphone or decrease the motion and/or quality of the video picture for better audio reception. Included is a network analyzer that displays the kilobit-per-second speed, the amount of packet loss, and the frame rate.

Personal Opinion

This one's the best of the bunch in terms of features, interface, clear video, and sound quality. And it's the only one that adheres to the true concept of a video phone. The resolution and clarity of the picture, even when someone is using a relatively cheap color camera, is amazing. And unlike FreeVue, which offers a rather amateurish interface, VDOPhone's interface is neat, clean, and very professionally designed, with all the options available at the click of a drop-down arrow.

I just got off the VDOPhone with a young man from Miami Beach, and our conversation was perfect. When I asked, "Shouldn't you be at work today? Are you playing hooky?" he laughed. He was using a black-and-white Connectix QuickCam, which is not considered high-quality by any means, but the picture was very clear. I could see him sitting in his room, with the sunlight shining through the curtains. Ah…to be in Miami in the winter…

He had turned off his microphone because he thought people couldn't hear him, so at first we communicated with the text chat option. When I asked if he had a microphone, he tried turning it on and soon his voice came through my computer speakers crystal clear. Almost perfect, in fact, with few or no break-ups. A nice feature of VDOPhone is that even with bad connections, you can adjust the quality of the video and the amount of refresh or motion you see to provide for better audio quality. And you can always increase the size of the picture to almost fill up your screen.

You connect to others either via an IP address or through the VDOPhone Web page listing, much like FreeVue. Instead of just a simple nickname list, you're also shown how long people have been on the net using VDOPhone, and whether they have a video camera connected to their computers.

Even if you don't have a video camera, you should try this program. You'll be amazed at the audio quality. One thing to remember, though—this program requires a relatively fast computer, so don't go complaining if your 386 can't run it. *Something* has to do the audio and video compressing. If you can find someone who does have a camera, you'll be amazed by the picture quality.

One other thing to remember: like all videophone programs, there are plenty of people who want to take their clothes off in front of you. And with VDOPhone you will certainly get a clear picture of them.

VidCall

MRA Associates
Home Page: http://www.access.digex.net/~vidcall/vidcall.html
Platform: Windows 3.1, Windows 95
Cost: Free
Rating: ☎☎

Minimum Requirements

- 486/66MHz PC
- 4MB RAM
- 2MB free disk space
- Compatible video capture card and camera, such as a QuickCam or VideoBlaster

Features

VidCall lets you send and receive scaleable video, text chat, transfer files, work with an interactive whiteboard, and share applications. You can send audio and video mail, collectively view images, and share applications. VidCall's Shared Workspace lets you bring in. bmp, .pcx, .gif, .tif, and .tga, files to work with collaboratively, and lets you include text from the Windows clipboard. Contents of the Workspace can be copied to the Windows clipboard for incorporation into other applications. Workspace tools include clip, crop, pencil, line, erase, and text, with variable font sizes and various colors.

You can use VidCall to connect to the VidCall server or contact someone directly via IP address. Multipoint video and document conferencing is available, accommodating up to ten participants (see Figure 10-4).

Personal Opinion

The concept that you can interactively share pictures, send audio mail, and see each other is fantastic, and VidCall's example of a doctor sharing an x-ray with another doctor over the Internet certainly gets you thinking about all the ways you could use interactive videoconferencing.

But other programs do all this and more, and they do it better and more reliably than VidCall. When I first tried the program, I didn't have a problem connecting to their demo server, but when I watched CNN, I couldn't hear anything. The problem? The default setting is no video, and for some Windows users, such as myself, I had to add additional GSM drivers in my Windows' system directory to make the program work.

I only discovered that after reading the help file. I believe a video phone program should sense that I have a sound card and install the appropriate drivers as the program is installed. Unfortunately VidCall did neither.

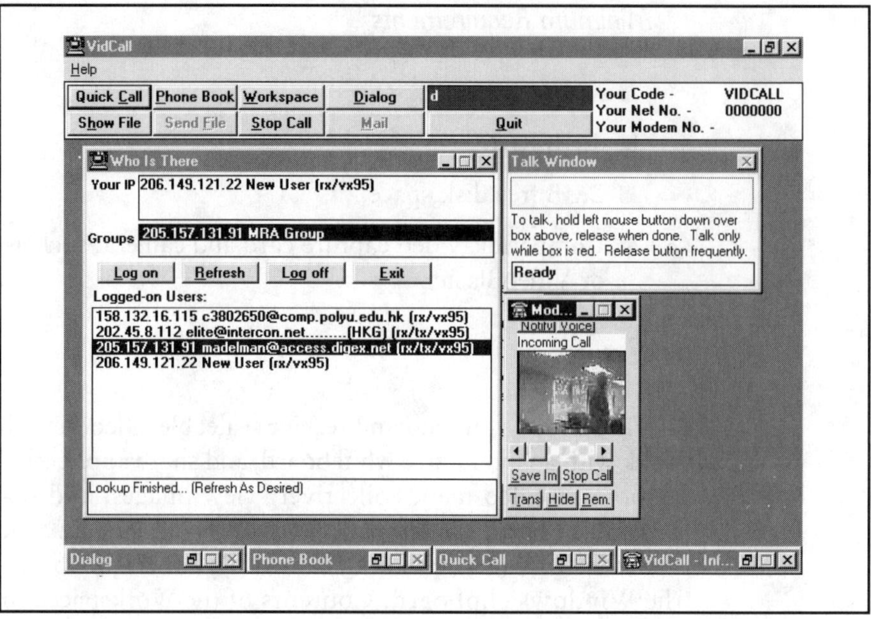

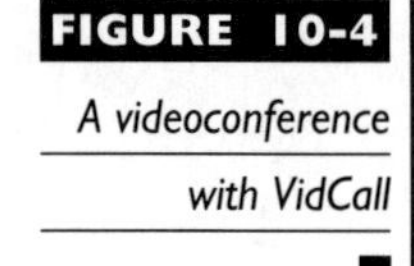
FIGURE 10-4

A videoconference with VidCall

The rest of my calls were hit-and-miss. I had no problem connecting with other people, typing messages, and sometimes seeing a rather grainy, often slow-to-refresh screen, but hearing voices clearly seemed next to impossible. I got nothing but choppy audio and loud feedback. The klunky and somewhat ugly interface was hard to understand at first, but once I read through the Readme a second time, I finally understood the function of some of the buttons.

VidCall is a program that needs a lot of tweaking to make it work. And you *really* have to read all the documentation before you even come close to a decent connection. It's not a program that is easy to use, or of the best quality.

Multiparty Conferencing Programs

PowWow

Tribal Voice
Home Page: http://www.tribal.com
Platform: Windows 3.1, Windows 95, Windows NT version 3.51
Cost: Free—Wow!
Rating:

remember *If you run a SLIP or PPP emulator such as SlipKnot, SLiRP, TIA, TwinSock, or Virtual TCP/IP, your computer will not have an IP address assigned to it, and other people using PowWow will be unable to reach you.*

Minimum Requirements

- 486DX/33MHz PC
- 4MB RAM
- 4MB free disk space
- Web browser (optional)

Features

PowWow (see Figure 10-5) is a multifunction conferencing program for the Internet that allows up to seven users to chat via text or voice, send files, view personal home pages and JPEG pictures, and cruise the World Wide Web (WWW) together. Version 2 added a multiple-user conference mode allowing up to 50 people to chat and cruise the Web together.

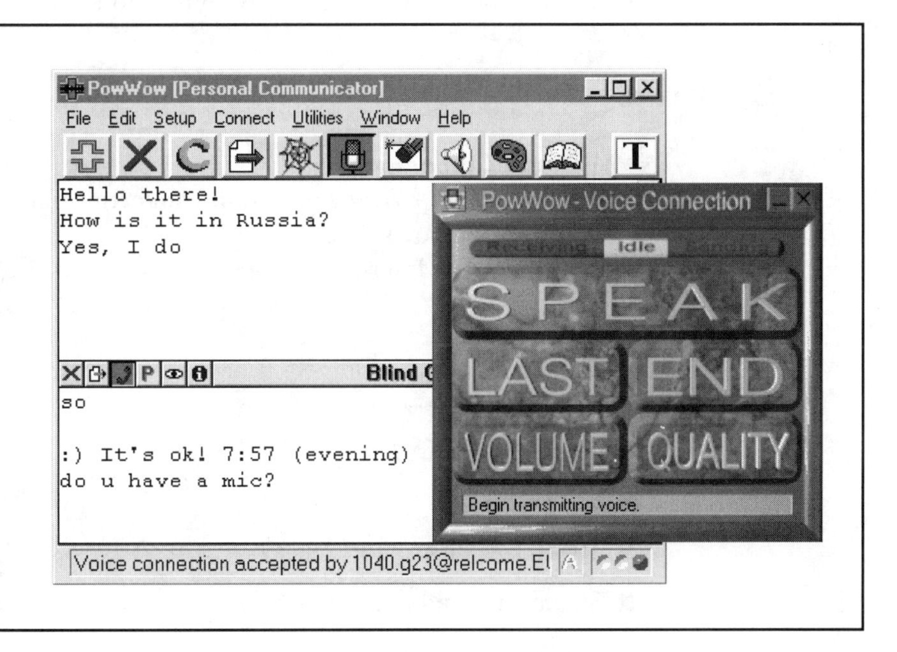

FIGURE 10-5 *Gathering around the conference table with PowWow*

PowWow is server-based, but the server never seems overloaded. You can chat, leave voicemail messages and connect to other users via the click of a Web page, or directly through an IP address.

Personal Opinion

Amazing. Free. Totally cool. Who could ask for anything more? This program rocks! These people really know what collaborative computing means. Everything *and* the kitchen sink.

I do have to tell you, however, that PowWow is not a program for the simple-minded. There are so many features that you should read all the documentation and FAQs before you fire up the program. And as with VDOPhone, if you want to use all the features of PowWow you should make sure your computer is powerful enough to run it.

You can add special sound effects to your text conferencing, and you can talk one-on-one to another person with the direct audio feature. The sound quality was excellent. Although you can't audioconference per se, you can point and click from one caller to the next, and invite them to an audio chat.

The neatest thing about this program is that people can take you on "tours" of Web sites. You simply follow along as they guide you through cyberspace. And you won't have to worry about finding other PowWowers to chat with; this is one of the most widely used and constantly updated conference programs available. And, again, it's free. What more could you ask for?

NetMeeting

Microsoft Corporation
Home Page: http://www.microsoft.com/netmeeting
Platform: Windows 95 only
Cost: Free with Internet Explorer (which is also free).
Rating: ☎☎☎☎☎

Minimum Requirements

- 486/66MHz PC
- 8MB RAM
- 10MB free disk space
- Sound card and microphone for voice communications

Features

Microsoft NetMeeting provides voice, text chat, whiteboard, application sharing, and file transfers (see Figure 10-6). Multiple users can share data, text, and applications in a single session, as you can see in Figure 10-6. You can call other NetMeeting users and voice chat with them, while you text conference with other users. You find other users to connect to via the User Location Service Directory, a server linking other NetMeeting users together, or you can connect directly via IP addresses. Even if you are already in a NetMeeting conference, other users can join in and drop out—or be booted out—of the conference at your direction.

Application sharing allows you and other NetMeeting callers to use a particular program on your or their PC. What you see on the screen can be seen on the others' screen, and you can give control of the program over to another caller. This means you can show people your spreadsheet figures, instruct others on how to create templates in Word, browse the Internet, or just play a game of Hearts. The person or people connecting to you don't need the shared application on their computers in order to see, or even run, the program. You can also use the file transfer option while you are

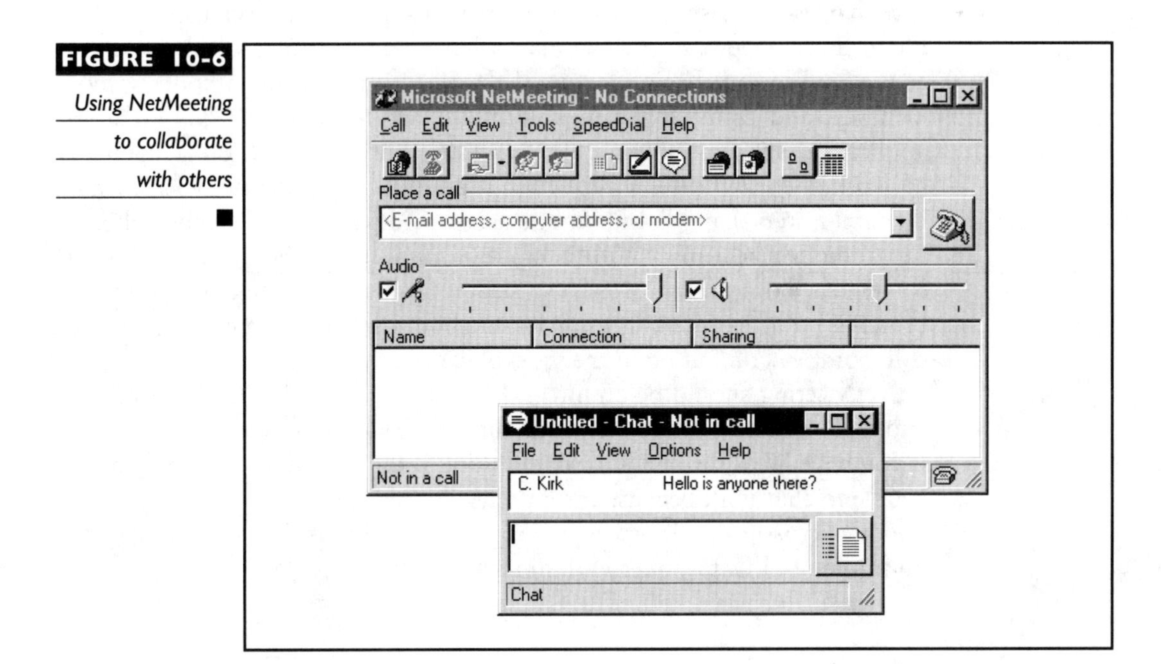

FIGURE 10-6 *Using NetMeeting to collaborate with others*

conferencing with other people to transfer computer files to conference attendees' computers.

There is also a shared clipboard, which makes it easy for you to copy and paste information from your computer's applications to NetMeeting so other users can see. You can copy and paste Web page addresses, PowerPoint slides, or Excel charts, just to name a few things. With the text chat feature of the program you and other conference members can type text, then save the conference to a file that can be posted to a Web page later.

Personal Opinion

You really can't beat NetMeeting for no-nonsense videoconferencing. Like PowWow, it offers everything you'd want in a conferencing program and more, but unlike PowWow, NetMeeting does it in a more staid, businesslike manner. When you get right down to it, the program's list of features is almost dizzying. You'd think that, with as many functions as this program has, you'd be overwhelmed by the interface, but Microsoft has created a very clean look and offers single pushbutton features that open into multiple windows.

The sound quality is really superb. I've yet to have a very choppy or inaudible conversation. The whiteboard feature is well thought-out, as is the text chat feature, especially since it lets you save your text chats into a text file for later review, something many chat programs don't allow. But as far as the application-sharing feature, I think more powerful computers, more memory, and faster Internet connections are needed. For dial-up users, application sharing simply isn't practical, but for those on a company network, the program is a must-have. Company computer networks, which operate faster than dial-up modems, provide the speed such elaborate programming requires. With a dial-up connection, the screen redraw is somewhat slow and can confuse people if you don't tell them you are bringing up a program. This program should certainly change the face of collaborative computing in the workplace.

NetMeeting should be an integral feature of upcoming Office products, and look for Microsoft to all but dominate the software conferencing arena. If you have Windows 95 it's worth taking it for a spin. There are plenty of people to chat with because it's a pretty popular program. You might even find a Microsoft developer or two to talk to. I did just last week and ended up trading Bill Gates jokes with him.

3-D OnLive! Traveler

OnLive! Technologies
Home Page: http://www.onlive.com
Platform: Windows 95
Cost: Free
Rating: ☎☎☎☎✆

Minimum Requirements

- Pentium
- 16MB RAM
- 13MB free disk space
- Sound card if you want to hear audio

note *A 586 chip other than Intel's, such as Cyrix's clone chip, may not work with this product. Also, SLIP is not currently supported, and neither are MWave sound cards found in some IBM computers.*

Features

OnLive! Traveler offers a truly unique conferencing experience by taking you to a virtual three-dimensional world (see Figure 10-7) where you can see other users ("avatars"), hear them speak, talk to them and freely move about, mingling with other travelers as you move through different worlds. You can use your own voice, or disguise it. OnLive! Traveler works as a stand-alone application or as a Web browser helper application, allowing you to visit different Web sites as you chat.

Multiple avatars can talk, not just chat, in real time. The further away you get from another avatar, the more distant his or her voice gets. The voice quality is excellent if you are using a fast enough computer and have a relatively clean Internet connection. You can pick from a large list of avatars or heads to represent you in the virtual world. All avatars are three-dimensional and can be customized with features you choose. You can add information about yourself to the properties of your avatar and disguise your avatar's voice if you prefer.

FIGURE 10-7 *Venturing deep into exotic virtual worlds with OnLive! Traveler*

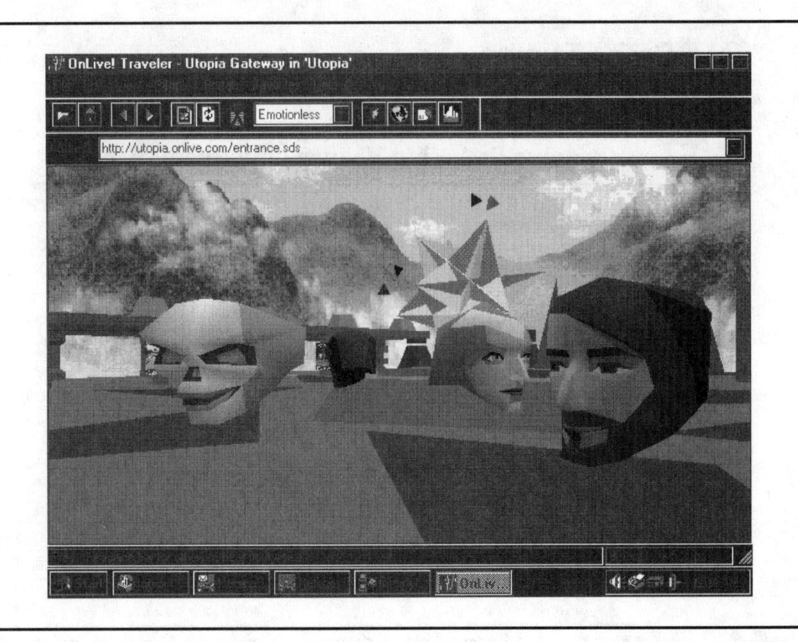

Look for an addition to OnLive! Traveler, called OnLive! Talker to be available soon. OnLive!'s newest application for voice conferencing on the Web turns normal Web sites into communities, letting people talk with each other right from their Web pages, without having to enter any specialized 3-D worlds. Talker will work with both Netscape Navigator and Microsoft Internet, and not only lets you talk to other people, but lets you e-mail them while you're in the Talker program.

Personal Opinion

Awesome! The first time I tried the program I connected with no problem and was amazed to hear several people talking—and clearly I might add. I must admit, I don't normally like virtual reality programs. But this one is not only delightful in its use of backgrounds, character graphics, and music, but also impressive in its ability to produce clear sound. I could actually understand and talk to other people. I felt like I was at a cocktail party, with clumps of people here and there chattering away. I'd move over to a group, introduce myself, and just start chatting. When I wanted to get a good look at someone, I would inch myself closer, often bumping into other people's heads. When I wanted to express my satisfaction or dissatisfaction with something, I could make funny faces with my avatar head.

When I got bored with one group conversation, I just moved my floating head toward another group of heads. As I moved closer, the sound of the people's voices got louder, and the sounds of the last group I had just left grew softer. As these talking heads moved to the left and right of me, so did their voices, cycling back and forth through the left and right speakers on my computer, and all around as they floated by.

The virtual portals or rooms are nicely designed. The sound effects are pretty darn funny. (When you run into another user, you hear this bonking noise, like you've hit a steel post.) The facial movements when people talk or express emotion are quick and well-designed. And when the rooms aren't too busy, you connect quickly and hear people very, very clearly.

Even if you don't want to audioconference, and if you're like me and hate virtual worlds, you should try this one. It's so cool, I had a hard time finishing this book because I was spending all my time bonking into other heads, talking about all sorts of interesting things with my virtual pals, and simply having a great time floating from one virtual world to the next. I started longing to see and talk to my newfound friends every afternoon.

OnLive! Traveler will definitely change your mind about how feasible virtual world multiperson voice conferencing can be. If I were giving out awards for the coolest program on the Internet, OnLive! Traveler would win top prize.

Chat Programs

NetPopup

JamSoft
Home Page: http://www.vtoy.fi/~malo/netpopup.html
Platform: Windows 3.1, Windows 95
Cost: Shareware. Under $25.
Rating: ☎☎☎☎

Minimum Requirements

- 486/33MHz PC
- 4MB RAM
- 2MB free disk space

Features

NetPopup is an Internet utility specifically for sending messages in real time (see Figure 10-8). You can type a message, click the Send button, and have that message appear immediately on the other NetPopup user's screen. No waiting for e-mail servers to deliver your message.

The Lookaround Seeker feature lets you browse for other NetPopup users within a certain IP range and quickly connect to others by simply clicking on their names. This is particularly useful if you are on an intranet and only want to search for people within your network.

You can even save, print, and forward your NetPopup text conversations. The small application runs minimized in your taskbar waiting for other NetPopup users to send you messages.

Personal Opinion

Even if you do have a multimedia-equipped PC, you should keep this little gem running all the time in your taskbar so people can send you quick messages, and you can respond instantly. It's better than e-mail because you are, in fact, communicating directly with the other person in real time. It's great if you have a dynamic IP address and want to let the other person know

FIGURE 10-8 *Chatting up a storm with NetPopup*

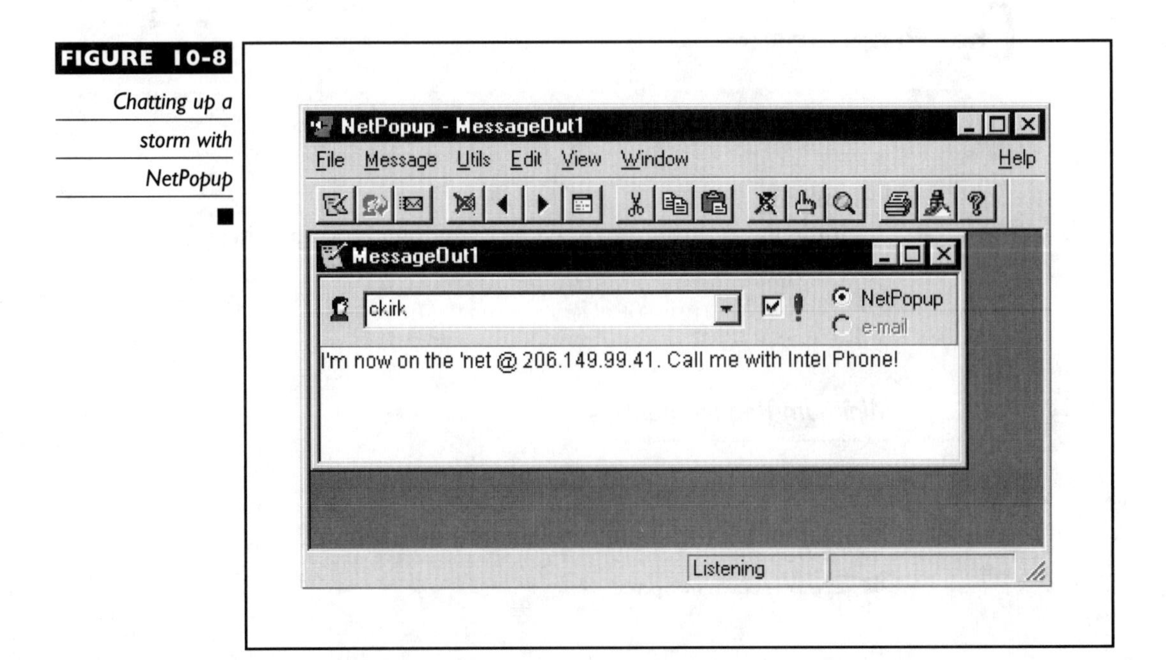

right away that you are online and ready to talk. It offers an IP Wizard that will not only get your IP address, but also e-mail it to your friends to let them know you are ready to talk.

The clean interface, along with handy little Tips of the Day make it very easy to use. It even comes with an Auto-Reply feature so you can alert people you might be away from your computer for a while but are still logged on to the Net. Plus you can cut, copy, and paste text to and from NetPopup, and the program even has an option to send your entire text chat conference to a particular e-mail address via Microsoft Exchange. If you are on an intranet or you have a dynamic IP address and don't want to rely on slow e-mail to let someone know you're ready to chat, this is the program to use.

WinPager

Rick McClanahan (author)
E-mail: rickm@hunterlab.com
Home Page: http://mason.gmu.edu/~rmcclana
Platform: Windows 3.1, Windows 95
Cost: Shareware. $5 registration.
Rating: ☎☎☎☎☎

Minimum Requirements

- Virtually any PC
- 4MB RAM
- 1MB free disk space

Features

WinPager (see Figure 10-9) lets you page another person over the Internet, as long as he or she is running WinPager too. You can also e-mail the person with your message at the same time that you WinPage him or her. Or you can establish a point-to-point text chat session with the person you are trying to talk to. WinPager can also notify you when the other person has received your message, letting you know immediately if the person is offline or not available to take your page.

Various sound effects notify you of different functions. WinPager will beep when you send a message or play back a voice file saying "message sent" when you click the Send button. Much like the ability to quote e-mail messages, WinPager can include the previous message in your WinPage

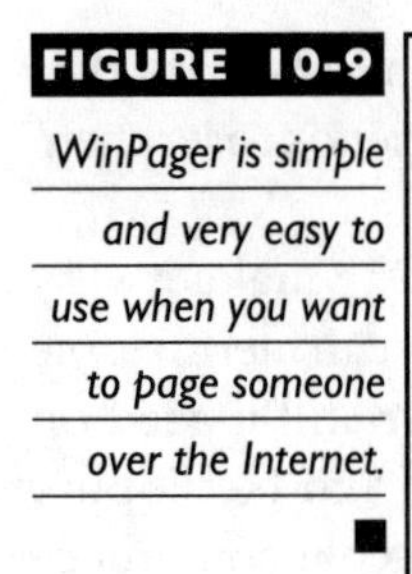

FIGURE 10-9

WinPager is simple and very easy to use when you want to page someone over the Internet.

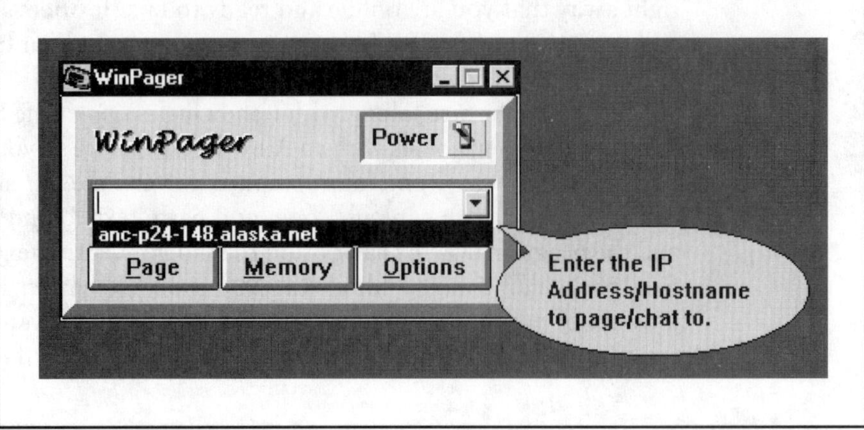

reply. You can also store people's names and e-mail addresses in a very simplified phone book.

Personal Opinion

Hey, we've got Internet phones, we've got virtual cocktail parties, we've got video phones. It was only a matter of time we got virtual pagers. This little program is well-designed, simple, and works flawlessly to quickly deliver text messages to another user. It works as a point-to-point communications tool, so you don't need to connect to a server. However, this program may not work very well if you have a dynamic IP address, because you need to know the IP address in order to send a page. However, if you check the TIPS section, you can find some handy tips on ways to find and post your IP address for all the world to see.

Give WinPager a whirl, if you don't like NetPopup. It only takes about two minutes to get up and running. The only downfall is that, like so many other quick-chat type programs, it isn't cross-platform compatible, which means I'll never be able to page my Mac friends with this product.

Internet-to-Phone Gateways

Net2Phone

IDT Communications Inc.
Home Page: http://www.net2phone.com
Platform: Windows 3.1, Windows 95
Cost: Free. (However, U.S. and international calls are billed at varying rates.)
Rating:

Minimum Requirements

- 486/33MHz PC
- 4MB RAM
- 4MB free disk space

note *At this time you cannot use Net2Phone behind a firewall.*

Features

Net2Phone connects your multimedia Internet-connected PC to a telephone network and lets you place long-distance calls (see Figure 10-10). The person you are calling hears the telephone ring, just as if the call were placed with a real telephone. The conversation is held via your microphone, sound card, and speaker combination.

Basically the Net2Phone software connects you to the IDT phone switches and converts the information from the Internet to the circuit-switched telephone network. The result is real-time uninterrupted voice communication between the two calling parties.

FIGURE 10-10 *Making a call with Net2Phone*

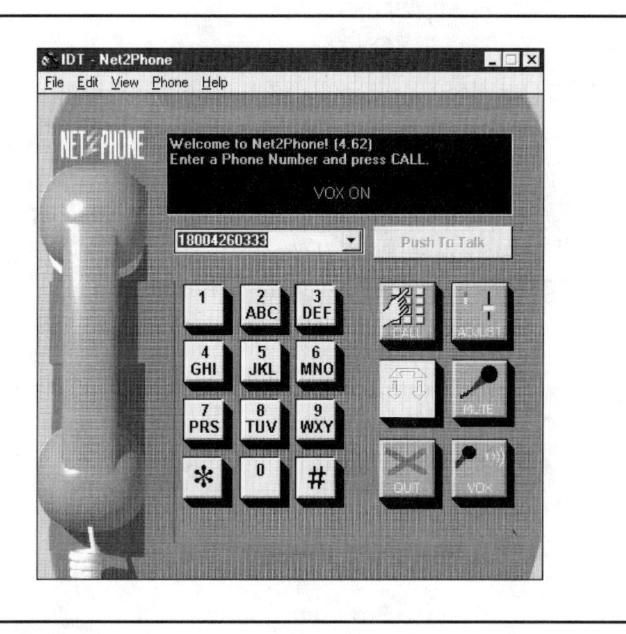

You can make unlimited 800 number calls, or use IDT's debit calling card to call any telephone number in the world for rates as low as ten cents a minute worldwide.

Personal Opinion

I hate to keep using the word "amazing," but that's the only word I can think of when describing this product. This program looks like a telephone (even more so than WebPhone). The first time I called the 800 number for Alaska Airlines flight arrival information to test Net2Phone, not only was I able to clearly hear what their automated voicemail system said, but I could also use the touch-tone keys on the telephone to pick and choose different options and navigate through their voicemail system.

You might be saying to yourself, "Well, heck, I can do that with my computer's telephone dialer program. What's so special about that?" What's so special is that long-distance phone calls overseas can be as low as ten cents. What long-distance phone company offers those kind of rates?

Better still, if you use your computer while you travel, you only need to make one phone call to check your e-mail, browse the Web, and dial out to other phone numbers without incurring expensive calling card and hotel charges. The party on the other end hears only a slight delay, and this type of delay is dependent upon the Internet traffic.

To see what I'm so excited about, I highly recommend you try this product, especially if you travel overseas and want to make long-distance calls to home.

GXC

Global Exchange Carrier
Home Page: http://www.gxc.com
Platform: Windows 3.1, Windows 95
Cost: Free. (However, U.S. and international calls are billed at varying rates.)
Rating: ☎☎☎

Minimum Requirements

- 486/25MHz PC
- 8MB RAM
- 4MB free disk space
- Internet Phone 3.2

Features

GXC lets anyone with a Windows-based multimedia computer call any telephone in the world (see Figure 10-11). You use the GXC software in conjunction with Internet Phone to place phone calls through the GXC telephone network; you then speak and listen to the other party via your own microphone and speakers. GXC provides its own discounted rates for long-distance calls.

The GXC software provides call tracking, discounted directory assistance to directory assistance operators all over the world, a personal directory or phone book with drag-and-drop features, and regular updates on the calling areas serviced by GXC.

Personal Opinion

I can't get too excited about GXC, because Net2Phone does such a better job of connecting PC/Internet users to the telephone. The GXC software does the dialing part okay, but then you have to use another program, Internet Phone, is to actually speak to the person on the other end. Having to bring up one program, then jump to another is way too complicated for most people.

In addition, the proposed rates at the time this book was written were only slightly less than the best long-distance rates most telephone companies

FIGURE 10-11 Using the GXC software to dial a call

were offering. My recommendation? Take it for a spin, but don't go outside your driveway with it.

Web-On-Call

NetPhonic Communications
Home Page: http://www.netphonic.com
Platform: Windows 3.1, Windows 95, a Sun SparcStation for the Web server, or even a telephone
Cost: Free to users. (However, server costs vary according to configuration.)
Rating: ☎☎☎☎

Minimum Requirement

- A telephone

note *This is not a program per se. It is a server-based product; therefore the user doesn't need anything but a telephone.*

Features

Web-On-Call marries the Web, the telephone, voice response, faxing, and information combining them all into a single server-based system. You create your company's Web pages. You put them on your Web-On-Call server (see Figure 10-12). Customers who have access to the Web can view your pages with any browser or they can use the Web-On-Call browsers to hear the embedded audio you've recorded.

Anyone can access Web pages anywhere without a computer or Internet connection. People who want to browse pages need only a touch-tone phone, cellular phone, or fax machine to reach information they want and have Web pages read, faxed, e-mailed, or sent to them by U.S. mail.

People without access to the Internet can use their touch-tone phones to listen to the same pages, navigating the links by pressing the # key on the touch-tone phone when they hear a prompt for a link they want to hear/go to. If they would like printed information they can request the Web pages be faxed or snail-mailed to them.

FIGURE 10-12

The Web-On-Call system

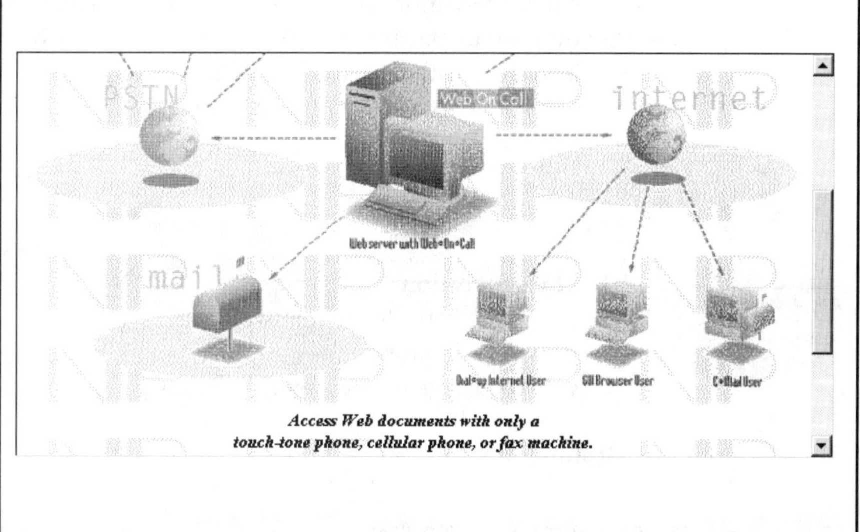

Web-On-Call basically is a fax-on-demand and voicemail system that reads Web pages. Eventually Web-On-Call will also be able to read e-mail messages directly from your mail server to you over the phone.

Personal Opinion

This like some of the other Internet-to-phone gateway programs is hard to explain, but has enormous potential for businesses that want a single site for company information. Just think. Although everyone seems to be on the Net, in reality more people have access to phones and fax machines than they do Internet browsers. Why spend thousands of dollars on a Web site if a large majority of the country can't access that information? By putting your pages on a Web-On-Call server, you have your fax-on-demand system, your voicemail system, and your Web site all in one place. You don't have to maintain separate systems, or worry about having some information updated on the Web site, but the information on the faxmail system is outdated. Your product catalog can be on your Web site, and can be made available via faxmail all at the same time, all in the same set of documents.

This is definitely more of a business than consumer product. But if you are thinking about putting in a Web server for your business, you should look at the Web-On-Call server before buying expensive hardware that may not provide you with as integrated a system as Web-On-Call.

DialWeb

Telet Corporation
Home Page: http://www.dialweb.com
Platform: Telephone
Cost: No charge for browsing; various pricing plans for storing audio files.
Rating: ☎☎☎☎☎

Minimum Requirements

- Web Browser
- Telephone, if you want to update your sound files remotely
- RealAudio Player

note *Like Web-On-Call, this product has two parts—the user part, and the server part. The requirements for the user are listed previously. If you would like to serve-up DialWeb documents, please check the Telet's home page for the server requirements.*

Features

DialWeb (see Figure 10-13) lets you update audio, graphics, or text on a Web site using only a touch-tone telephone. DialWeb combines RealAudio with an integrated voicemail system that listens to the responses given via a touch-tone phone and implements them into the specific Web page.

There is no special software needed to update the information. Audio files and the voicemail system are run through Telet directly. The user never has to download any program other than a standard Web browser to see and hear the changes made over the phone.

FIGURE 10-13

Dialing for Web pages with DialWeb

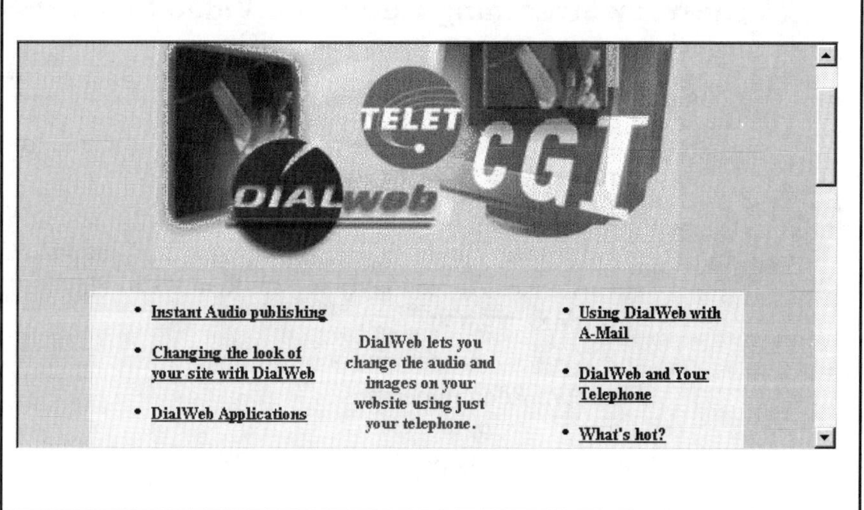

Personal Opinion

This is a great concept. Your company's Web site could include a daily audio "briefing," that you can quickly and easily update anywhere there is a phone. No need to be at the computer or in front of a microphone. Just use your touch-tone phone and your voice to update the information.

You can call in weather reports, sports scores, price updates, or whatever else your imagination dreams up. One of the demonstrations DialWeb uses is that of a weather page. You use your touch-tone phone to update this page, punching in the high and low temperatures, pick from a list of weather options such as partly cloudy, cloudy, or sunny, and record a voice message with tomorrow's forecast. All this is done by using the keys on your touch-tone phone to input the highs and lows, select the type of graphics to resemble the cloud coverage, and have your voice recorded in RealAudio format. Then anyone with Web browser can see and hear the changes almost instantaneously.

If you want to or have plans to add audio to your Web pages but don't want the hassle of having to maintain a server, invest in recording equipment, and be tied to your Web server, DialWeb is the answer.

One-Way Streaming Audio and Video Products

RealAudio Player and RealAudio Player Plus

Progressive Networks
Home Page: http://www.realaudio.com
Platform: Windows 3.1, Windows NT, Windows 95, Macintosh
Cost: RealAudio Player is free; RealAudio Player Plus is $29.95.
Rating: ☎☎☎☎☎

Minimum Requirements

- 486/25MHz PC
- 8MB RAM
- 4MB free disk space

or

- 68040 Macintosh
- MacOS System 7.1
- 8MB RAM
- 4MB free disk space

Features

RealAudio Player lets you listen to live and prerecorded audio files immediately without having to download the entire audio file. This streaming audio feature means you can listen to live broadcasts of radio stations, TV stations' audio, and live broadcasts over the Internet.

RealAudio Player comes as either a separate application or as a Netscape plug-in, and offers various user-adjustable controls (see Figure 10-14), including volume control, program fast-forward or rewind (available only with previously recorded audio files), and Start and Stop buttons. RealAudio Player also alerts the user to the name of the audio file, the length, how many minutes or hours of the file that have been played, and the author of the audio file.

Hundreds of radio stations around the world use RealAudio to broadcast their programs. Many more such as ABC News, NPR, and the ESPN Radio

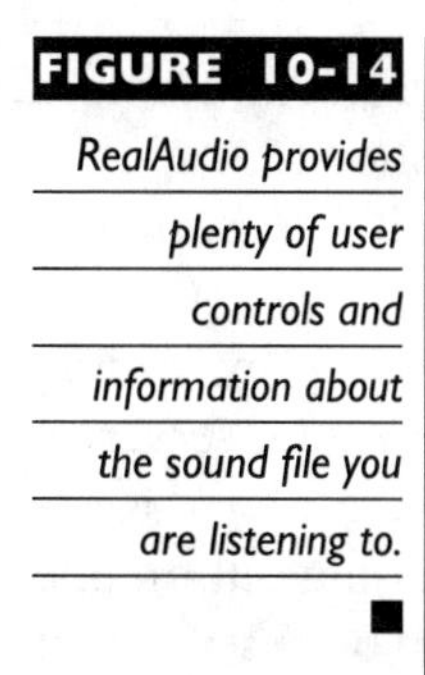

FIGURE 10-14 RealAudio provides plenty of user controls and information about the sound file you are listening to.

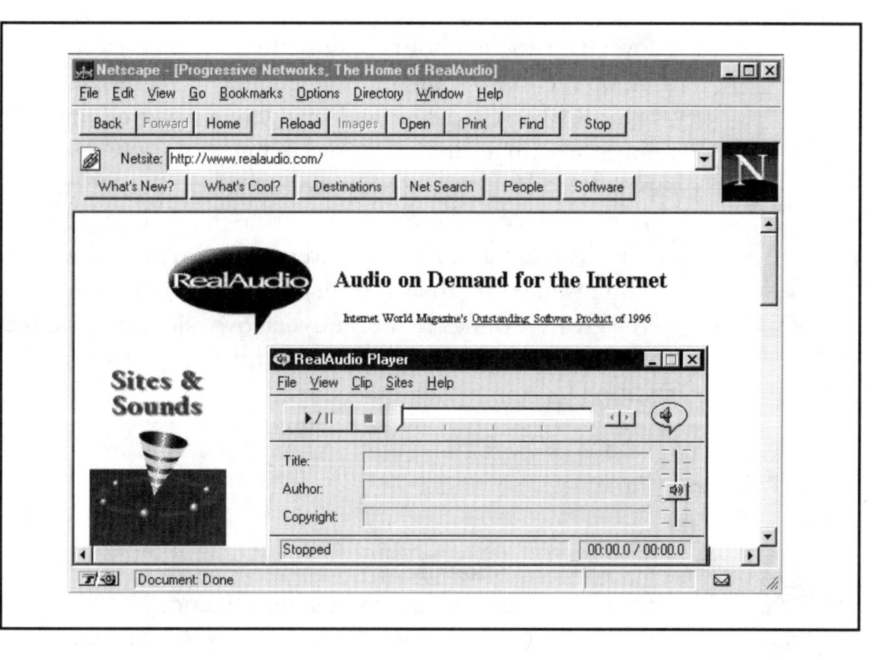

Network save various live radio programs so that listeners can hear entire programs any time they want.

RealAudio Player Plus lets you scan the hundreds of radio stations as you would a regular radio, without having to go to each individual radio station Web site to connect to the station. RAP also lets you program buttons for your favorite RealAudio sites, so you can click a button and have RAP take you to the RealAudio enabled Web site. For sites that offer recorded information, RAP lets you download the entire RealAudio file and save it to your hard disk so you can listen to it at any time without having to be connected to the site, as with the standard RealAudio player.

Personal Opinion

Who wouldn't love RealAudio? This program brings the world to your desktop. The sound quality is usually very clear and infrequently choppy, and best of all, often *live*. I can sit in my office in Anchorage listening to a radio station broadcasting from Hong Kong without fiddling with short-wave radios, or putting up a huge antenna or paying for a satellite dish.

If I miss one of my favorite talk shows, like Gina Smith's On Computers, I can go to her Web site, click a RealAudio link and listen to her Sunday call-in talk show, even though it's Tuesday. And since it's streaming audio,

I don't have to wait hours to download a single 15-minute segment. I simply click on the one I want and immediately start hearing the show.

Plus RealAudio can provide me with a never-ending supply of new music, right at my fingertips, without having to pay a dime for expensive CDs or tapes. It truly brings excitement to often dull Web browsing by adding extremely clear audio to any Web site that is using the RealAudio server.

This is truly a must-have Internet tool. Anyone who doesn't have Real-Audio installed on their Internet connected computer is missing out on one of the greatest things to hit the cybertown since the graphical browser. If you don't have it, get it now. You're missing out on a lot. Remember, real audiophiles don't download.

StreamWorks

Xing Technologies
Home Page: http://www.xingtech.com
Platform: Windows 3.1, Windows 95, Macintosh
Cost: Free
Rating: ☎☎☎☎

Minimum Requirements

- 486/25MHz PC
- 4MB RAM
- 4MB free hard disk space

or

- 68040 processor
- MacOS System 7.0
- Sound Manager 3.0 or higher
- 8MB RAM
- 4MB free disk space

Features

StreamWorks plays live and on-demand audio and video from Stream-Works servers across the globe. You don't have to wait to download the video or audio file. Much like RealAudio, StreamWorks provides an immediate and continuous stream to your computer.

StreamWorks is based on the industry standard MPEG format, the international standard for digital audio and video for a multimedia PC. There is no special hardware needed outside the standard video monitor and sound card.

The StreamWorks player (see Figure 10-15) provides full-screen, full-color, full-motion video with CD-quality (44KHz) audio, depending upon the speed of your Internet connection.

If you have a slow connection, StreamWorks can scale its transmission of huge files, from 2.0 megabits down to a relatively small file, such as 8.5 kilobits so the file can be delivered over 14.4 Kbps and 28.8 Kbps modems. This means an entire newscast could be compressed into a relatively fast, small streaming file you could watch on your PC.

The StreamWorks 2.0 player also lets you adjust the quality of the audio and video. You can choose to listen to audio only, basically using StreamWorks much like RealAudio to listen to live radio-type broadcasts, or you can turn off the sound and simply view the video. You can always adjust the data connection.

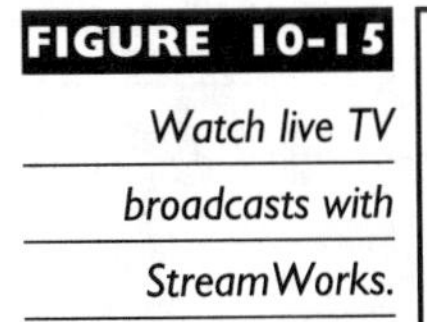

FIGURE 10-15 *Watch live TV broadcasts with StreamWorks.*

Personal Opinion

StreamWorks' quality is relatively good, although not quite up to TV refresh-rate standards, but the audio never seems to be compromised by the video. You can always clearly hear the audio portion of the program with little or no breakup.

I think the program is fabulous. I can watch MTV without having to buy cable. But the interface on the PC version is just not as clean as other streaming video products, such as VDOLive. The Mac version, however, seems much more streamlined and refined.

StreamWorks does have one of the most diverse collections available of live and recorded video files, so you are sure to find plenty to watch and listen to.

VDOLive

VDOLive
Home Page: http://www.vdolive.com
Platform: Windows 3.1, Windows 95, Macintosh
Cost: Free
Rating: ☎☎☎☎✆

Minimum Requirements

- 496/66MHz PC
- 8MB RAM
- 4MB free disk space

or

- 68040 Macintosh
- MacOS System 7 or higher
- 8MB RAM
- 4MB free disk space

Features

The VDOLive video player provides real-time playback of video files over the Internet through 14.4 and 28.8 Kbps modem connections. The user can adjust and control the size of the picture, the volume, and the quality of audio and video delivered (see Figure 10-16).

Web designers embed VDOLive documents as plug-ins for Netscape Navigator users, or as ActiveX controls for Internet Explorer 3.0. The browsing user see an icon representing the embedded movie and can click it to play the film or to control the file.

Personal Opinion

The video quality and sound quality is top-notch, and the refresh rate one of the best in the live audio players. The fact that you can embed VDOLive files into a Web page means you can bring to life your Web pages with full-motion video and audio.

I think VDOLive offers much better streaming video quality than StreamWorks, with a cleaner interface. But you really should try both since the type of content offered by various radio stations, TV stations, and private individuals varies greatly between products. The main problem with both programs is that heavy Internet traffic sometimes slows the video refresh rate to a crawl.

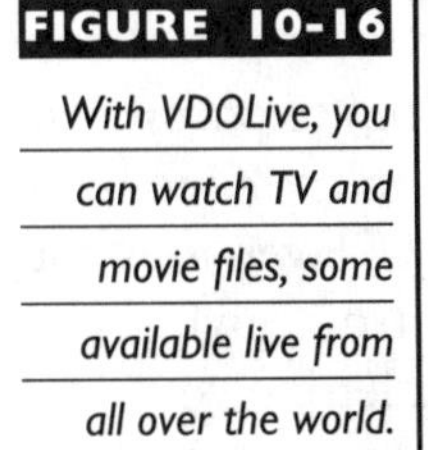

FIGURE 10-16 With VDOLive, you can watch TV and movie files, some available live from all over the world.

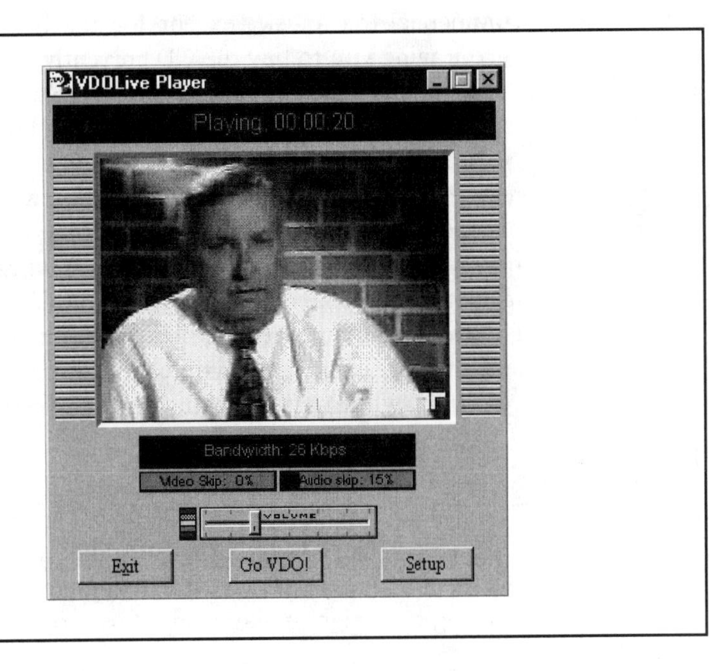

The DJ Player

Terraflex Data Systems Inc.
Home Page: http://www.thedj.com
Platform: Windows 95, Windows NT
Cost: Free
Rating: ☎☎☎☎✆

Minimum Requirements

- 486/66MHz PC
- 8MB RAM
- 2MB free disk space
- RealAudio player

Features

The DJ Player offers you quick and easy access to music with over 40 different channels or types of music to choose from, via a single program that can occupy your screen or minimize and continue to play. The DJ Player plays music 24 hours a day with the help of the RealAudio player. It's commercial-free music, except for the banner that periodically appears, encouraging you to buy the CD currently playing.

Essentially, The DJ Player utilizes the RealAudio technology to play music (see Figure 10-17). But when you use The DJ Player, you don't *need* to have Netscape or RealAudio up and running. Instead, The DJ Player uses the features of RealAudio, but provides you with its own controls, giving you one-click access to over 40 different types, or channels, and letting you click from one music selection to the next. You can preset the program to your favorite music channels and have them play the minute you launch The DJ Player. Since the program is self-updating you get the latest music immediately, free of charge, and you can minimize the program, moving it out of your way, while music streams through your computer speakers.

Personal Opinion

What a perfect way to use RealAudio. Instead of having to move from one Web site to the next, it's all there in one simple package. When you get tired of listening to Dwight Yokum sing those cowboy blues, simply click the Show

FIGURE 10-17 *Listen to any kind of music with The DJ Player and RealAudio.*

Tunes button, and you're listening to a relatively clear rendition of "Oklahoma." Click another button to hear the latest Goo Goo Dolls song.

The DJ Player really turns your computer into an unlimited juke box. What could be better? My folks use it all the time to listen to show tunes, and they love it. It's entirely free, supported by people who buy the CDs advertised by The DJ Player. If you're like me and you're too cheap to buy music, give this product a try.

Appendix A

Instructions for Using the CD-ROM

NOW that you've read all about Internet telephones in this book, you're ready to try out some of the programs. The Internet Phone Connection CD-ROM contains plenty of programs to choose from. Some are limited in the number of minutes or days you can use them; others let you talk as long as you like. I've designed the CD to give you a taste of Internet phone technology. It will very likely get you hooked on using the Net to talk to friends, relatives, and business associates.

What Will You Find on the CD?

Throughout this book you've seen the CD-ROM icon next to descriptions of various products. This icon means that the program and its manufacturer's home page can be found on the CD. For other products, I've at least included the manufacturers' home pages on the CD so you can easily get more information directly from the source. All you have to do is bring up your favorite browser, then open the home page for the manufacturer you want.

The CD was written to be used in either a Mac or a PC with at least a 2× CD-ROM drive. If you use a Mac or a Windows 95 computer, you'll notice that the filenames are truncated in order to comply with DOS naming conventions which require filenames of no more than eight characters with three-letter extensions. The Mac programs have already been decompressed, and are ready to be installed once you've copied them to your hard drive. For Windows 3.1 and Windows 95, you'll need to decompress some of the files. I've included the WinZip decompression program to make this task easier.

When you first load the CD, you'll see four folders, or directories. Let me explain what each directory holds.

The Help Folder

The Help folder contains an interactive guide to installing both the Mac and PC phones. You don't need to run the interactive Help program in order to install the programs, but if you're a novice, I'd highly recommend it. In order to use the Help file, your computer needs to meet the following requirements:

Minimum Requirements: PC

486/33MHz PC
8MB RAM
9MB free disk space
Sound card and speakers
Apple QuickTime 2.1
Monitor capable of displaying 256 colors

note *If you need to install version 2.1 of QuickTime, you can download it from* ***http://quicktime.apple.com****. The Help file will not work with previous versions of QuickTime.*

Minimum Requirements: Mac

68040 Mac
8MB RAM
9MB free disk space
Sound Manager 3.0
Apple QuickTime 2.1 or above
Monitor capable of displaying 256 colors

for mac users *A Macintosh with 4MB of real RAM memory doubled to 8MB with RAM Doubler will not operate.*

Installing the Help File

Installation for PCs

The Help file should run without a problem directly from the CD on your PC, as long as your PC has met the minimum requirements. First, make sure you hae set your monitor to display at least 256 colors through the Control Panel/Monitor settings. If your monitor isn't capable of displaying 256 colors, you will not be able to run the Help file. Also, make sure no applications are running when you start the Help file.

Open the Help folder and find the file entitled "Program" and then double-click it. In a few minutes a screen will appear, and you will hear my voice welcoming you to the installation process. If you don't hear my voice, make sure your speaker volume is turned up. If the program reports that you must have 256 colors set, check your Control Panel settings to ensure your monitor is set correctly.

If a message appears on your screen stating that you must use a newer version of Apple's QuickTime, make sure you download that from **http://quicktime.apple.com**.

If you receive any other error message, please send me e-mail at **netphones@aol.com**, and I will try to resolve your problem. It wasn't possible for me to test the CD on every possible computer configuration, so you may run into problems that relate specifically to your particular setup, which I may not have encountered when I made the CD. I'll do my best to help resolve any problems you might have, but again, you don't need to run the Help file in order to use the programs.

Installation for Macs

In order for the Mac version of the Help file to work you must first copy the Help folder—just the Help folder, not the other folders—to your hard disk. In order to do this, you must have at least 9MB of free hard disk space available. Once you've run the Help file, you can then delete it from your hard drive.

Highlight the entire Help folder and drag it to your computer's hard disk. Do not drag it to a folder inside your hard drive; instead copy it to the main level of your hard drive. It should take a few minutes for the files to copy. Once finished copying, open your hard drive, and rename the Help folder **NETPHONES**. (Make sure you've named it "NETPHONES" in all upper-case letters.)

Before you run the Help file, make sure you have your monitor set to display at least 256 colors. To do this, go to the Apple/Control Panel settings and look for the Monitor icon. Set the levels of colors to 256 before you start the Help program.

Also make sure no programs are currently running, and that you have rebuilt your desktop prior to running the Help program. This will prevent many common installation problems. To rebuild the desktop, restart your Macintosh and hold down the COMMAND (or Apple) key and the OPTION key on your keyboard. Keep holding those two keys down as your Macintosh starts. You can let go of those keys once you see a dialog box starting with "Are you sure you want to rebuild the desktop on...". After the computer has finished rebuilding the desktop, restart your Mac, and then locate the NETPHONES folder.

Next, open the NETPHONES folder. You should see several files with the word HELP in them. Find the file labeled "_HELP_" and double-click it. In a few minutes, your screen will go entirely black, then show the main Help screen. If the main screen doesn't appear within a few minutes, you may be experiencing one of the following problems:

- You don't have QuickTime 2.1 or above installed.
- You have not rebuilt or restarted your Macintosh.
- You are using RAM Doubler or another memory-enhancing program.
- You do not have enough RAM (at least 8MB) to run the program.
- You do not have your monitor set to 256 colors.
- You did not double-click on the _HELP_ file.
- You may be experiencing problems with an INIT, Extension or Control Panel that is conflicting with QuickTime and/or the Help program.

In order to resolve conflicts with other programs, make sure you are not running any other programs, and that you have restarted your Mac. If the Help program still won't run, try turning off some of your extensions with Extension Manager, or through the Control Panel settings.

If you are still having problems, feel free to e-mail me at **netphones@aol.com.** It was impossible to test this program on every single type of configuration, but I might be able to help you with your particular system. And remember that you don't need to run the Help file in order to use the programs.

The Others Folder

The Others folder contains other types of Internet phone-related software, such as video phones, directory assistance programs, and Internet-to-phone software, which are described in Chapter 10.

When you first open this folder, you'll see two subdirectories inside. The Mac directory contains the videoconferencing and live streaming audio and video products specifically for the Mac, and the PC directory holds the same types of programs made specifically for the PC. Make sure you read the minimum hardware requirements for these programs before you install them. Many of these programs require hefty computers to run.

The Phones Folder

The Phones folder, just like the Others folder, is divided into PC and Mac subdirectories. Each of these holds numerous Internet phone programs for each type of machine. I've tried to give these folders names that are as close to the actual names of each of the Internet phone programs as possible, to make it easier for you to find the Internet phone program you want to try.

Some of the phone directories however, contain only a text file, which is actually the manufacturer's home page. In order to use this file you must first bring up your Web browser, then choose File/Open and point the computer to the directory of the Web page file you want to open. Once opened, you will be able to download and/or register the phone program you want to use.

I also recommend you check all the manufacturer's Web sites periodically for news and updates on their products, even if the phone program you want to use is included on the CD. And don't forget to check my home page at **http://www.netphones.com** for additional technical support and updates.

The WinZip Folder

The WinZip folder is only for PC users, and is used to unzip or uncompress those files that are compressed with the WinZip program.

To install WinZip on your computer, locate the version for your particular operating system:

WINZIP95.exe—WinZip for Windows 95 users
WZ16V6.exe—WinZip for Windows 3.1*x* users

Double-click the appropriate file for your computer, and the installation process should start, prompting you for the location where you want to store WinZip. If you find that you use WinZip regularly to uncompress other programs, you should register and pay for WinZip, one of the best shareware programs ever made. You can do this by visiting the WinZip home page, which is included on the CD in the same WinZip folder.

Installing the Programs

For PC Products

To install the program you want, locate it from the CD and copy it to the TPhones folder you created in Chapter 10. Then double-click on the file you've just copied, and it should either self-extract or bring up the WinZip program. Follow the instructions on the screen for installing the program, or refer to the interactive Help program for more information.

For Macintosh Products

To install the Macintosh program you want, locate it on the CD and copy it to your hard disk. Then open the newly copied folder found on your hard drive and double-click the program icon. Depending upon the program you are installing, the program should immediately start, or take you to a configuration screen. If the program asks you for a registration number, be sure to visit the manufacturer's Web site for this information. (Sorry, I don't give out registration numbers.)

Problems? Questions? Answers?

Again, feel free to e-mail me at **netphones@aol.com**. If you are having problems, please make sure you include the following information:

- The type of computer you use—Mac or PC
- The model number of the computer
- The amount of RAM memory and hard disk space
- The version of the operating system you are using
- The type of sound card you are using
- A concise description of the problem

I check e-mail regularly and will try to answer your questions as quickly as I can. You might also check any readme files, and my Web site at **http://www.netphones.com** for additional information. Also, you can join the Netphones discussion group by sending me a blank e-mail message. In the subject line, put "subscribe" and your first and last name, and I'll add you to the list.

Good luck, and I'll talk to you on the Net!

Glossary

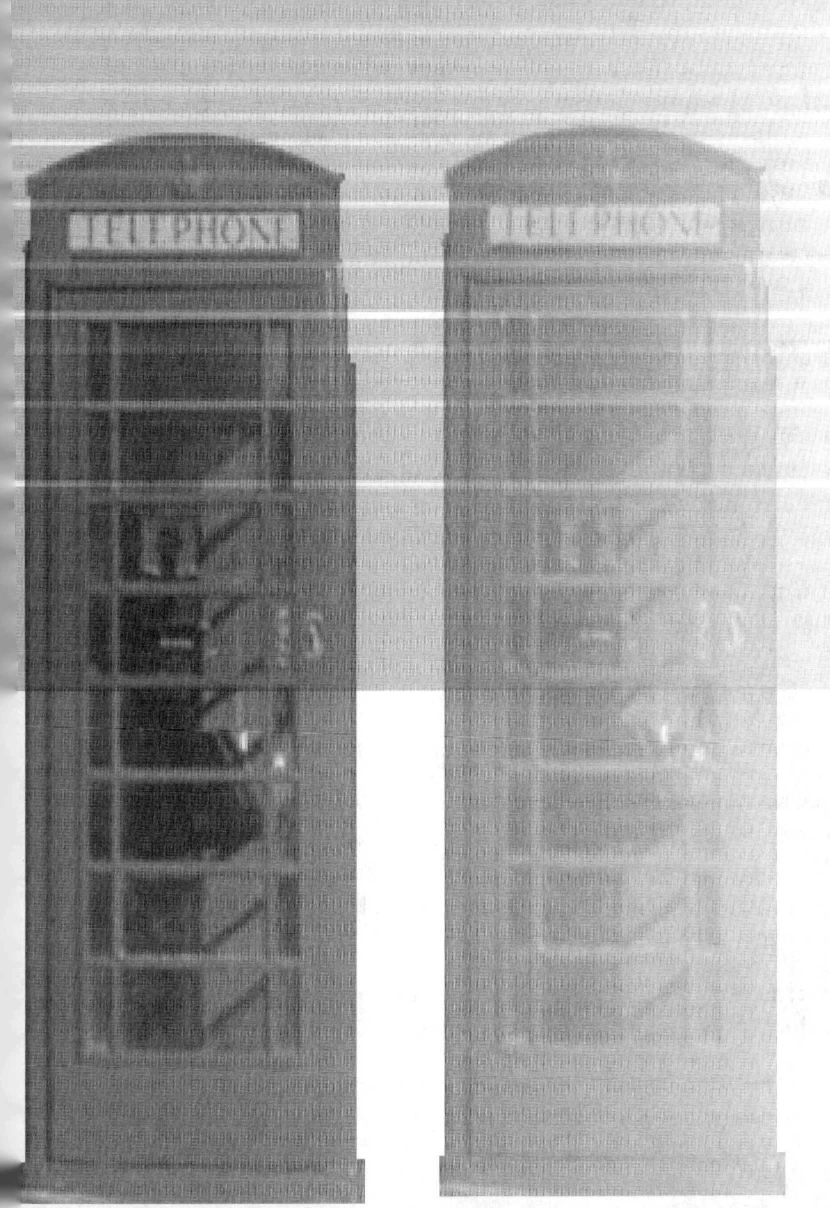

ADPCM (ADAPTIVE DIFFERENTIAL PULSE CODE MODULATION) A type of compression that halves the data rate to 4000 bytes per second. ADPCM's resulting compression is identical to simple compression, but the fidelity loss is much less. ADPCM requires a relatively fast computer. ADPCM encoding of sound data requires much less storage space than PCM, the format used by WAV and AIFF files.

AIFF (AUDIO INTERCHANGE FILE FORMAT) Developed by Apple Computer, AIFF is an audio file format used to store high-quality audio and musical data. It is a common format on both Macintoshes and PCs.

AM (AMPLITUDE MODULATION) A method of broadcasting in which the desired audio or video signal modulates the amplitude of a "carrier" signal.

AT COMMAND SET A type of language, i.e. list of commands, that controls a modem. Hayes Microcomputer Products developed this language for their modems to understand how to do tasks such as dial the phone, adjust the speaker volume, or have the modem answer an incoming call.

ANALOG Information that is reproduced using a continuously varying electronic signal. (Compare with *digital*.)

ANONYMOUS FTP A method for transferring files via an FTP server. Anonymous FTP lets you log in to an FTP server as an unregistered user. Your user name is anonymous and your password is your e-mail address.

ANSI (AMERICAN NATIONAL STANDARDS INSTITUTE) An organization of American industry groups that works with the standards committees of other countries to develop universal standards to promote international trade and telecommunications. ANSI has helped promote such standards as ASCII, SCSI, and the Ansi.sys device driver.

AOL (AMERICA ONLINE) The biggest consumer online information service offering a graphical communications interface for PC and Mac users alike.

API (APPLICATION PROGRAMMING INTERFACE) An API is a series of functions that various programs can utilize to make the operating system perform standard tasks such as opening windows, displaying message boxes, or handling files. Oftentimes single instructions are used to pass information from the program to the operating system. Windows has several APIs that deal with telephony, messaging, and database issues.

ASCII (AMERICAN STANDARD CODE FOR INFORMATION INTERCHANGE) Pronounced "askee," ASCII is a standard developed by the American National Standards Institute (ANSI) that defines how computers read and write characters. The 128-character ASCII set includes numbers, letters, punctuation, and control codes. Each of these characters is represented by a number. Most operating systems, excluding Windows NT, use the ASCII character set.

AU A type of audio file format, most commonly used on Unix computers. However, a Macintosh or PC running the right program, such as the Netscape browser, can play AU files.

AVI (AUDIO/VIDEO INTERLEAVE) A type of video file format, used by Windows, in which the picture and sound elements are stored in alternate, interleaved chunks within the file. It is one of three different types of video file formats used on personal computers. MPEG and QuickTime are the others.

BANDWIDTH The amount of data capacity a network can handle. Oftentimes people refer to it as in "I need more bandwidth to get a good clean Internet phone connection." This would mean the person may need a faster connection to the Internet. Bandwidth is measured in cycles per second or hertz (Hz), or the difference between the highest and lowest frequencies transmitted, or by measuring the bits or bytes transmitted per second.

BASS The lowest audible sound frequencies, between approximately 20 and 200 cycles per second.

BAUD A measure of how frequently a signal changes per second. High modem speeds really should be specified in bits of data transferred per second, not baud. You'll see many modem manufacturers use the term *baud* when they mean *bits per second,* or *bps.*

BETA The second iteration of software before the final release of the product. When something is in "beta testing," it is almost complete and ready for commercial sale, but still requires people to test it and report bugs or problems with the software to the manufacturer.

BIT (BINARY DIGIT) A bit is the smallest unit of data measured in computers. The value can be either 0 or 1. It can also be used to describe data transfer speeds in modems, as in *bits per second.*

BROWSER A program that lets you read World Wide Web documents. Companies that produce browsers are NCSA, Netscape, Microsoft, CompuServe, Prodigy, and America Online, just to name a few.

BYTE A unit of measurement of data. A byte contains eight bits.

CACHE A portion of RAM or the hard disk that stores information accessed over and over again. Used to speed up computers and Web browsers by storing information such as graphics in a directory that can be accessed quickly.

CD (COMPACT DISC) A digitally encoded record, aluminized to reflect light and played with a low-power laser. A 12-centimeter (4.7-inch) CD contains up to 74 minutes of stereo sound.

CD-ROM (COMPACT DISC-READ ONLY MEMORY) A laser-read, computer-accessible disc that stores large quantities of digitized data.

CLIENT Either the computer, the software, or the user accessing data from a central computer or server. Your Netscape browser is the client that accesses information on the Web server.

CODEC (COMPRESSION/DECOMPRESSION) An algorithm or mathematical equation that reduces audio file sizes. Codecs convert the analog sound or video into digital signals on the sending end and decode the sound and video into analog sounds on the receiving end. When using it with your Internet telephone, both parties have to use the same codec. Different codecs work best with different types of machines and different bandwidth or Internet connection speeds. Some codecs include GSM (Global Standard for Mobile Communications), a codec used in cellular telephones in Europe, ADPCM, PCM, and DSP TrueSpeech.

COM PORT Short for communication port, which refers to your PC serial port. On a PC you may have up to four or five different communication ports, plugs that communicate with modems or other computers. They are usually named COM1, COM2, etc. COM is also synonymous with serial port.

COMPUSERVE One of the oldest online commercial service providers, offering dial up access in over 100 countries, and throughout the US.

CPU (CENTRAL PROCESSING UNIT) The main engine or heart of your computer. The CPU does all the computing, and in the case of Internet phones, is responsible for taking instructions from the codec to compress and decompress the sound.

CRT (CATHODE RAY TUBE) The screen usually used on a TV set or on a computer terminal.

CYBERSPACE An electronic world you can get sucked into. Very few ever return.

DATA PACKET A chunk of computer data organized and labeled for transmission on a network or serial line. The TCP/IP protocol used on the Internet sends around 1500 characters in a single packet. Standard modem-to-modem communication sends approximately 64 characters per packet.

DIGITAL A method of representing signals as a series of binary numbers. (Compare with *analog*.)

DISTORTION Changes in a signal that involve the addition of spurious tones at frequencies not present in the original sound, or noise in the conversation that affects the quality of the conversation.

DNS (DOMAIN NAME SERVER) An Internet server that keeps track of and matches IP addresses with names. For example, the computer running my Web server software is identified by a unique number, 206.149.99.41, but to make it easier for people to find, it's been given a name of www.anchorage.org. The DNS server matches the real number of the computer to the name given to that number.

DROPOUT A momentary loss of sound in a conversation, usually caused by a slow computer processing a codec meant for a more powerful computer, or by heavy Internet traffic.

DSP (DIGITAL SIGNAL PROCESSING) A special-purpose microprocessor that handles signal-processing programs very quickly. DSPs are found in many modems and sound cards, and offer real-time compression of sound and video, thus taking some load off your computer.

DUPLEX The ability to talk and listen (transmit and receive) at the same time. Half-duplex refers to only being able to do one thing at a time, either listen or talk, but not both. Full-duplex lets you perform both functions at the same time. Speakerphones are commonly half-duplex devices, as are many of the sound cards found in today's computers.

DYNAMIC RANGE The ratio between the loudest and softest sounds that can be reproduced accurately by a recording medium.

E-MAIL (ELECTRONIC MAIL) Messages that are transmitted to and from your computer and another computer and specified for a particular recipient. E-mail is usually stored on an e-mail server until the recipient connects to the server computer and requests his or her incoming mail.

ECHO BACK To play so the user can hear and evaluate. Speak Freely offers an echo server, a computer that records your voice and sends it back to you, replaying it through your speakers. Your voice and the voice of the person you're talking to can also echo back through your speakers as you speak. This is caused by the other party having his or her speakers and microphone too close to each other.

EQUALIZATION Changes in amplification at low or high frequencies; used to compensate for the limitations of a recording medium and to obtain equally accurate reproduction at all frequencies.

FAQS (FREQUENTLY ASKED QUESTIONS) Usually, a text file that lists frequently asked questions and answers about a particular topic or product. Make sure you check out the FAQs for each of the Internet phone programs you use. You'll always find valuable information about the product and how compatible it is with various computers, browsers, and Internet service providers.

FAT BINARY A file larger than the original application, which provides program code for several types of computers, such as PowerMacs and regular Macs.

FAX (FACSIMILE TRANSMISSION) A system that electronically transmits and receives text over ordinary phone lines.

FILE EXTENSION Usually a three-letter designation after the name of the file and preceded by a period that tells you and the computer what kind of program created that file. Generally found on DOS-based computers, but also on the Internet to designate the file or compression technique used, such as .bin for binary, .sit for Stuffit compressed files, or .zip for WinZip or PKZIP compressed files.

FINGER A program you use to find out either a list of users currently logged on to a system, or information about a particular user on a system. When you finger someone's Internet account you should see the person's full name, most recent log-in, and the contents of their Plan file if they have one. Most systems, however, don't let you finger from outside their network.

FIREWALL Usually a dedicated computer equipped with security programs that check both incoming and outgoing data transmissions. Firewalls protect networks from access by non-authorized users from outside the network. Firewalls also are used to block access to certain types of Web sites and data transmissions.

FM (FREQUENCY MODULATION) A method of broadcasting or recording in which the desired audio or video signal modulates (varies) the frequency of a "carrier" signal.

FREEWARE Software you can download and pass around to friends without cost or signing a license. The only caveat is that freeware is still copyrighted, meaning you can't charge for it, since you don't own the rights to the program. Freeware was originally a trademark of Andrew Fluegelman, one of the inventors of shareware and freeware. Freeware is rarely supported by the person who created the program.

FREQUENCY The number of cycles per second of a transmission. One hertz (Hz) = 1 cycle per second, 1 kilohertz (KHz) = 1000, 1 megahertz (MHz) = 1,000,000, and 1 gigahertz (GHz) = 1 billion.

FTP (FILE TRANSFER PROTOCOL) A protocol or way in which you can copy files between your computer and another computer. FTP is a client/server-type setup, offering an FTP server that multiple people can access and download files from. You can also access FTP sites via your browser by typing **ftp://*username:password@domain*** in the Location field of your browser, replacing *username* with your login name, *password* with your login password, and *domain* with the name of the FTP server you are trying to access.

FULL-DUPLEX Having the ability to speak and listen (receive and transmit) at the same time. Duplex and full-duplex mean the same when you are talking about sound cards and Internet telephones.

GATEWAY A program or piece of hardware that passes data between different types of networks.

GIGABYTE A unit of measurement for space on a hard disk or removable data cartridge. A gigabyte is 1,073,741,824 bytes, and is usually abbreviated by the letters GB. You may need a gigabyte of space to hold all these Internet phone programs.

GSM (GROUPE SPECIALE MOBILE, OR GLOBAL STANDARD FOR MOBILE COMMUNICATIONS) A type of codec or sound compression algorithm used in European cellular telephones that uses a fairly powerful computer to compress sound at a high-quality rate.

HALF-DUPLEX The transmission of data in only one direction at a time, resulting in the limitation of not being able to speak and hear at the same time. Many of today's computer sound cards are half-duplex, which means that you have to click on a button to talk and then usually click on that same button again to listen (and to alert the other party you are in listening mode).

HAYES-COMPATIBLE Refers to whether a modem understands the AT command sets, or language, found in Hayes modems.

HELPER APPLICATIONS Additional programs that add other features to your Web browser, such as the ability to show QuickTime movies, or play sound files.

HTML (HYPERTEXT MARKUP LANGUAGE) A set of standard instructions used to format documents for displaying on the World Wide Web.

IMPEDANCE Opposition or resistance to the flow of electrical current. The rated impedance of a loudspeaker is an average, since the impedance depends on the frequency of the signal.

INTERNET BACKBONE The main "pipes" that connect major metropolitan areas to one another. Comprised of national Internet service providers (ISPs) such as Net 99 and Alternet. These ISPs offer T3 links running at 45 Mbps. Your local ISP most likely connects to this backbone through their routers and phone lines.

IP (INTERNET PROTOCOL) The common language spoken among all computers on the Internet so that each can pass data back and forth.

IP ADDRESS The number assigned to your computer once you log on to the Internet or once your network is connected to the Internet. Each computer and device on the Internet must have a unique address or number that identifies it to the rest of the Net. An IP address is a unique string of numbers, shown in groups and separated by periods, such as 204.17.139.1. The first several numbers help to identify the geographic location of the equipment and the last two sets help identify the network and particular machine being used.

IRC (INTERNET RELAY CHAT) A way to connect people on the Internet together in such a way as to offer real-time chatting or voice communications. IRC works with servers and clients. You connect to a server and use a client to talk to or text chat with other people connected to that same server. IRC servers oftentimes have multiple channels or categories to help people decide what topic they may want to discuss. VocalTec's Internet Phone and SoftFone use the IRC technology to link callers together.

ISP (INTERNET SERVICE PROVIDER) The company that gives you access to the Internet. Local ISPs are usually a bunch of Jolt cola-drinking genius types who like to play with Unix and routers all day. These lovable geeks are responsible for keeping Web servers, mail servers and FTP servers, among other things, up and running.

ISDN (INTEGRATED SERVICES DIGITAL NETWORK) A way of connecting multiple types of telephone equipment into a single line. ISDN offers digital instead of analog connections, but still uses existing copper telephone wires to transmit voice and data on the same line, and at high speeds, starting at 64 kilobits of data per second. ISDN connections may be fast, but they require special hardware, such as ISDN modems, and special configurations at the telephone company's end.

JAVA An object-oriented programming language that works with a wide variety of computers. Created by Sun Microsystems, Java adds animation and interactivity to Web pages, granted you have a Java-enabled browser.

KBPS (KILOBITS PER SECOND) A measurement of speed, referring to how much information can be transferred per second, measured in bits per second.

KILOBYTE A unit of measurement for data. "Kilo" is Greek for thousand, so *kilobyte* means a thousand bytes, but in actuality represents 1024 bytes.

LAN (LOCAL AREA NETWORK) A bunch of computers, printers, and servers connected together via cabling and network interface cards or plugs.

LEASED LINE A dedicated telephone line, used to tie networks (LANs) together. Leased lines transmit data at only one speed, and that speed is determined by the amount of bandwidth a customer purchases. Customers pay a flat monthly rate for this type of service.

LED (LIGHT-EMITTING DIODE) Solid-state devices that glow when electric current is applied. Also refers to little lights that blink in some Internet phone programs to let you know that another person is calling or is on hold.

LOSSLES A type of compression that throws away redundant bits of information without affecting the quality of the sound or image.

LOSSLY A type of compression that reduces the size of a file but, in the process, reduces the quality of the sound or image.

LOUDNESS COMPENSATION A tone-control process that boosts low frequencies at low volume levels in an attempt to compensate for the ear's insensitivity to quiet bass sounds.

LPC (LINEAR PREDICTIVE CODING) A compression scheme that reduces the data rate by a factor of twelve. LPC offers the greatest amount of compression, but requires a very fast computer.

MEGABYTE A unit of measurement for data. "Mega" is Greek for a million, but a megabyte actually contains 1,048,576 bytes or 1024 kilobytes. Oftentimes abbreviated as MB.

MIDRANGE Frequencies in the range spanned by the human voice, from approximately 200 to 2000 cycles per second.

MIME (MULTIPURPOSE INTERNET MAIL EXTENSION) The Multipurpose Internet Mail Extension helps various Internet applications, such as e-mail and World Wide Web browsers, understand and display different file formats. When a browser or e-mail program includes a particular Mime format, or when you add a new Mime type to it, the program decodes or interrupts the file and either launches the appropriate helper application or, if programmed to do so, displays the graphic or plays the sound from within the browser.

MODEM A device used for accessing computer data over telephone lines. *Modem* is short for *modulator/demodulator,* which basically means that the modem on the sending end is turning data signals into sound signals that can be sent over telephone lines, and the modem on the receiving end is turning the sound signals back into data signals.

MONITOR A video display designed to receive direct input of video signals from your computer. The thing that connects to your system and kinda looks like a small TV, which you look at when you use your computer.

MONOPHONIC OR MONAURAL Recording or playback involving only one channel of sound.

MULTICASTING The act of sending out data, either video or audio, to anyone who cares to tune in. You can multicast with Speak Freely and RealAudio, or when you conference call with several people using NetSpeak's WebPhone or VocalTec's Internet Phone.

NARROWCASTING Transmission to a specific, limited audience (such as Japanese-speaking people, for example), often via streaming products such as Real Audio or VDOLive.

ODBC (OPEN DATABASE CONNECTIVITY) A set of programming instructions that interface with different applications and define how to move information in and out of PC databases and spreadsheets that support ODBC.

OEM (ORIGINAL EQUIPMENT MANUFACTURER) A company that makes a product but allows it to be marketed under another company's brand name.

PIXEL (PICTURE ELEMENT) The smallest area of a video picture capable of being delineated by an electrical signal. The number of pixels in a complete picture determines the amount of detail or resolution in the picture.

PCM (PULSE CODE MODULATION) A method of digitizing analog sound so that it can be stored and played back on a computer. PCM is the most common technique for digitizing sound. It is used in WAV and AIFF files.

POP (POINT OF PRESENCE) A collection of modems, leased lines, and multiprotocol routers that allow local dial-up access, usually to national or commercial Internet Service providers, which in turn provide you access to the Internet.

POWER AMPLIFIER The portion of an amplifier that produces the high current levels needed to drive a loudspeaker.

PPP (POINT-TO-POINT PROTOCOL) The standard protocol or language used to communicate with the Internet via standard serial communications. PPP defines how your modem connection will exchange data packets with other computers and devices on the Internet.

PROTOCOL A collection of rules that computers have to follow in order to communicate with each other. TCP/IP, PPP, and SLIP are protocols, as are HTTP and FTP.

QUADRAPHONIC Surround-sound reproduction involving the recording and playback of four channels of sound.

QUICKTIME Developed by Apple Computer, QuickTime is a method for storing sound, graphics, and movies. QuickTime files are designated by the .mov extension. QuickTime compresses multimedia information into small files that can be used on Macintoshes or PCs.

RAM (RANDOM ACCESS MEMORY) The working, temporary storage area of your computer where information and programs are placed as you use them. Think of RAM as the top of your desk and hard disk space as your file cabinet. When you want to work with a file, you pull it out of the file cabinet and place it on your desktop (RAM). As you jump from answering the phone to updating that file to writing things down on a legal pad, you are accessing the top of your desk (again, RAM). The more RAM you have in your computer, the faster it will work (less shuffling of papers), and the better your voice will sound over an Internet phone.

RESOLUTION A standard measurement for the amount of detail that can be seen on a computer screen, expressed in the number of horizontal lines on a test pattern.

RESONANCE The natural tendency of a device to vibrate at a specific frequency. Unwanted resonances in speakers, for example, alter the sound by producing excessive response at some frequencies.

RJ-11 The standard telephone connector plug at the end of your telephone line that plugs into the wall jack or modem plug. RJ-11 cabling uses two of four wires within the telephone line to transfer sound or data.

ROUTER A piece of hardware that routes information (data) to the appropriate place within a network. Routers act as traffic cops, making sure things get where they should be going and that only authorized machines transmit data into the local network. Routers also handle errors, keep network usage statistics, and offer a certain level of security.

RS232 (RECOMMENDED STANDARD 232) A standard type of plug used on the back of computers to connect to modems and other external devices. This type of plug connects to a cable through twenty-five pins lined up in two rows. One pin is designated as the ground, several others are dedicated to carrier signals, and still others are designated to keep the sending and receiving of data in synch.

SAMPLING Sampling is measuring the curves sound makes at regular intervals. Sound is basically a series of curved waves. Sampling a sound gives a rough, stair-stepped approximation of a curve. How well the sound matches the original is controlled by how often the measurement is taken (sampling rate), and how precisely the measurement is recorded (sampling resolution).

SAMPLING RATE Digital audio is recorded by storing samples. A sample has a size (16 bits for CD audio) and a rate (44.1KHz for CD audio). The higher the sampling rate, or the more often you record and measure the sound curves, the more data in the sample, thus the better the sound quality.

SERVER A computer whose sole purpose is to provide information to clients based upon the type of protocol they specify. Servers can serve up files, Web pages, voice, a list of Internet telephone users, or e-mail.

SHAREWARE Software that offers a "try before you buy" approach. You can download shareware, use it for a certain period of time, and if it works for you, pay the shareware author a nominal fee, usually less than $50.

SMTP (SIMPLE MAIL TRANSFER PROTOCOL) The protocol that regulates how mail servers communicate with each other and exchange e-mail. You need to know your SMTP mail server name to use products like DigiPhone, WebPhone, and CoolTalk.

S/N RATIO (SIGNAL-TO-NOISE RATIO) The range, usually expressed in decibels, between the loudest sound a recording medium can accommodate and its background noise level.

STEREOPHONIC Sound reproduction that uses two or more channels in order to represent the size or spatial distribution of sound sources.

SURROUND Sound reproduction that surrounds the listener with sound, as in quadraphonic recording and reproduction.

T1 A name coined by AT&T for a type of system/bandwidth that digitizes signals at 1.544 megabits per second. If you have a T1, you're cruisin,' baby.

T3 Thirty times the capacity of a T1 line, a T3 line can handle 44.736 megabits of digital data per second. Most large ISPs use T3 connections.

TCP/IP Transfer Control Protocol/Internet Protocol is a standard set of instructions that lets different types of computers "talk" or transfer data to each other. TCP/IP takes the information to be transmitted and breaks it into pieces. Each piece is given a number which the intended computer uses to verify and put back the data in it's proper order. The data being transferred via the TCP/IP protocol is placed into a TCP "envelope" which, in turn, is placed into an IP packet. When the receiving computer gets the envelope it takes the data out of the envelope and places it in the order specified. If data is missing or garbled, TCP automatically asks the sender to send the envelope again.

TELEPHONY Da bidness of telephones.

TEXT CHAT BOARD A screen that allows two users to type text messages to each other. TeleVox, FreeTel, and Internet Phone are just a few of the Internet phone programs that use text chat boards.

TREBLE The highest audible sound frequencies, between approximately 2000 and 20,000 cycles per second.

TRUESPEECH A codec or sound compression scheme of up to 8000Hz, in mono, making it one of the smallest sound compression schemes available. TrueSpeech sound quality isn't as good as a CD's, but more like that of a telephone.

TWEETER A small loudspeaker that reproduces high frequencies.

V.32 A modem standard for correcting errors and compressing data at 9600 bits per second.

V.32 VBIS A modem standard for correcting errors and compressing data at 14,400 bits per second (14.4 Kbps).

V.34 A modem standard for correcting errors and compressing data at 28,800 bits per second (28.8 Kbps).

V.34 VBIS A modem standard for correcting errors and compressing data at 34,400 bits per second (34.4 Kbps).

VDT (VIDEO DISPLAY TERMINAL) Your computer's monitor, or screen.

VIRTUAL MEMORY Part of a computer's hard drive that acts like memory. If you have virtual memory set up on your computer, when your RAM fills up, virtual memory kicks in and holds the information overflow. Oftentimes you'll hear the term *swap file* used with virtual memory. This is the file virtual memory uses and goes back to when swapping data between RAM and your hard drive. Having virtual memory turned on will slow down your computer and cause choppiness in your conversations.

WATT A unit of electrical or acoustical power. Electrical power is the product of voltage and current. Acoustical power is measured according to the intensity of sound pressure.

.WAV The standard extension for waveform sound files in Windows.

WHITEBOARD A special screen where you and the other parties can draw, type, or collaborate on ideas. NetMeeting, ClearPhone, and Internet Phone offer whiteboard features.

WINSOCK A piece of software that bridges your computer and the Internet. Winsock acts as a liaison between your Windows Internet programs, such as your Web browser and the Internet protocol. The Winsock instructions are stored in a .dll file.

WOOFER A large speaker that reproduces low frequencies—or a big dog that barks a lot.

Index

D

E

Z

$10 # COUPON $10

Upgrade to VidCall®
Standard Version and Take
$10.00 off the Regular Price of $55.00

Fax this coupon along with the Software Registration Form found in your demo software on enclosed CD-RM to MRA Associates, Inc. to receive your discount.

MRA Associates, Inc.

2102B Gallows Road
Vienna, Virginia 22182-3960
703-448-5373
Fax: 703-734-9825
E-mail: bleuthy@mnsinc.com

$10 $10

You are here, you are there and, you are ... there! You can Gather and Talk.

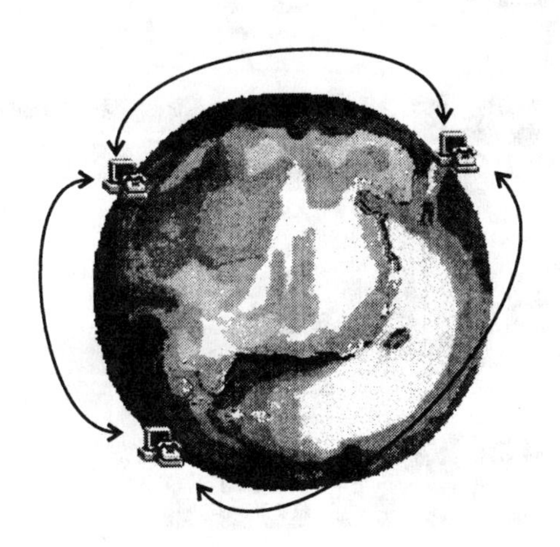

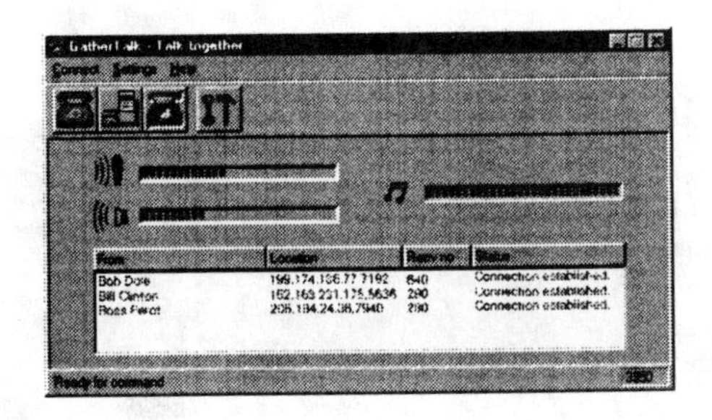

GatherTalk Features :

- true conferencing - supports up to 5 people using 28.8kbps modems
- runs on Win95 and Solaris
- exploits CPU fully to produce high quality voice
- actively developing with new features

http://www.cixt.cuhk.edu.hk/gtalk/

Centre for Internet Exchange Technologies
Tel (+852) 2609 8445 Fax (+852) 2603 5032
Email info@cixt.cuhk.edu.hk

INTRODUCING
Windows® 95 Internet & Sound Utility
Midisoft
Control mute, volume, balance, bass and treble
Create and send Audio Mail™ over the Internet
Record and mix sound with Audio Recorder
Locate, list and play all of your sound files with Sound Finder™
Record and store Audio Note™ reminders
Midisoft
SOUND BAR
Record great sound! Play it, add it to a document, or send it anywhere over your network or Internet e-mail.
CD-ROM
Includes A $10.95 Labtec Microphone!
FREE AT&T WorldNet Software Inside!
AT&T
Includes 2 Copies! One For You And One For A Friend!
Only $29.95!
INTRODUCTORY OFFER!
Get started right away with the included high-quality Labtec® AM-32 condenser microphone. A $10.95 value!
AM-32
Send Audio Mail
0:15
0:30
0:45
Subject My First Audio Mail
Auto Record Level
Compress Audio Mail
Auto Voice Detect
Return Receipt
For a FREE
30-Day Trial
visit our Website:
http://www.midisoft.com

WARNING: BEFORE OPENING THE DISC PACKAGE, CAREFULLY READ THE TERMS AND CONDITIONS OF THE FOLLOWING COPYRIGHT STATEMENT AND LIMITED CD-ROM WARRANTY.

Copyright Statement

This software is protected by both United States copyright law and international copyright treaty provision. Except as noted in the contents of the CD-ROM, you must treat this software just like a book. However, you may copy it into a computer to be used and you may make archival copies of the software for the sole purpose of backing up the software and protecting your investment from loss. By saying, "just like a book," The McGraw-Hill Companies, Inc. ("Osborne/McGraw-Hill") means, for example, that this software may be used by any number of people and may be freely moved from one computer location to another, so long as there is no possibility of its being used at one location or on one computer while it is being used at another. Just as a book cannot be read by two different people in two different places at the same time, neither can the software be used by two different people in two different places at the same time.

Limited Warranty

Osborne/McGraw-Hill warrants the physical compact disc enclosed herein to be free of defects in materials and workmanship for a period of sixty days from the purchase date. If the CD included in your book has defects in materials or workmanship, please call McGraw-Hill at 1-800-217-0059, 9:00 A.M. to 5:00 P.M., Monday through Friday, Eastern Standard Time, and McGraw-Hill will replace the defective disc.

The entire and exclusive liability and remedy for breach of this Limited Warranty shall be limited to replacement of the defective disc, and shall not include or extend to any claim for or right to cover any other damages, including but not limited to, loss of profit, data, or use of the software, or special incidental or consequential damages, or other similar claims, even if Osborne/McGraw-Hill has been specifically advised of the possibility of such damages. In no event will Osborne/McGraw-Hill's liability for any damages to you or any other person ever exceed the lower of the suggested list price or actual price paid for the license to use the software, regardless of any form of the claim.

OSBORNE/McGRAW-HILL SPECIFICALLY DISCLAIMS ALL OTHER WARRANTIES, EXPRESS OR IMPLIED, INCLUDING BUT NOT LIMITED TO, ANY IMPLIED WARRANTY OF MERCHANTABILITY OR FITNESS FOR A PARTICULAR PURPOSE. Specifically, Osborne/McGraw-Hill makes no representation or warranty that the software is fit for any particular purpose, and any implied warranty of merchantability is limited to the sixty-day duration of the Limited Warranty covering the physical disc only (and not the software), and is otherwise expressly and specifically disclaimed.

This limited warranty gives you specific legal rights; you may have others which may vary from state to state. Some states do not allow the exclusion of incidental or consequential damages, or the limitation on how long an implied warranty lasts, so some of the above may not apply to you.

This agreement constitutes the entire agreement between the parties relating to use of the Product. The terms of any purchase order shall have no effect on the terms of this Agreement. Failure of Osborne/McGraw-Hill to insist at any time on strict compliance with this Agreement shall not constitute a waiver of any rights under this Agreement. This Agreement shall be construed and governed in accordance with the laws of New York. If any provision of this Agreement is held to be contrary to law, that provision will be enforced to the maximum extent permissible, and the remaining provisions will remain in force and effect.

NO TECHNICAL SUPPORT IS PROVIDED WITH THIS CD-ROM.